Getting Started with DraftSight

Learn mechanical drafting for commercial
and academic projects

João Santos

PUBLISHING

BIRMINGHAM - MUMBAI

Getting Started with DraftSight

First published: August 2013

Production Reference: 1200813

Published by Packt Publishing Ltd.
Livery Place
35 Livery Street
Birmingham B3 2PB, UK.

ISBN 978-1-78216-024-3

www.packtpub.com

Cover Image by Sandeep Babu (sandyjb@gmail.com)

Credits

Author
João Santos

Reviewers
Deepak Gupta

Mark Lyons

Neb Radojkovic

Acquisition Editor
Mary Nadar

Commissioning Editor
Sruthi Kutty

Technical Editors
Vrinda Nitesh Bhosale

Mrunmayee Patil

Ruchita Bhansali

Project Coordinator
Michelle Quadros

Proofreader
Samantha Lyon

Indexer
Monica Ajmera Mehta

Production Coordinator
Melwyn D'sa

Cover Work
Melwyn D'sa

About the Author

João Santos is the manager and main instructor at QualiCAD (www.qualicad. com), one of the most important Portuguese ATCs (Autodesk Authorized Training Center), based in Lisbon. With a degree in Mechanical Engineering, he has been teaching AutoCAD for more than 25 years and, most recently teaching several AutoCAD clones such as DraftSight, ZW CAD, ProgeCAD, BitCAD. He is an AutoCAD 2013 and 3ds Max 2013 Certified professional user and instructor, being also the Portuguese instructor in these technologies with more students. João is the author and co-author of more than 40 Portuguese AutoCAD and 3ds Max books. He is also the author of *Autodesk AutoCAD 2013 Practical 3D Drafting and Design*, *Packt Publishing*.

First of all, I would like to thank my family for all their support and guidance. No less important are all my friends, students, readers and colleagues for their continuous questions, feedback, suggestions, and basically shaping my career. And I would also like to express my gratitude to everyone at Packt for this opportunity and collaboration.

About the Reviewers

Deepak Gupta graduated in 2000 from Indo Swiss Training Centre, (Chandigarh, India). With over 12 years of rich experience working with various industries, he has been working in different roles with different product lines. His main area of experience and interest has been the design and manufacturing processes along with team and project management.

He is working as Technical Director at Vishnu Design Services, an Engineering and CAD outsourcing service provider based in Chandigarh, India, offering wide range of CAD design, drafting, and engineering services. Being a certified SolidWorks expert and user for the last 7 years, Deepak is passionate about working with SolidWorks. Besides working for Vishnu Design Services, he enjoys writing for his own blog, *Boxer's SolidWorks Blog* where he writes tips, tricks, tutorials, and news about SolidWorks. In addition to this, he has participated in various SolidWorks World Conferences as a press member.

An important aspect of his life is his family—his parents, his wife, a 5-month old son, and other family members. He loves to travel and make friends.

I would like to thank my wife, Swati Gupta, for her love, kindness, and support she has shown during the past few weeks it has taken me to review this book. Furthermore I would also like to thank my parents for their endless love and support. I would also like to thank João Santos for writing such a great book, which I enjoyed reading more than simply reviewing. Last but not least, I would like to thank the Packt Publishing team for choosing me as one of the reviewers of this book and helping me out to complete the review of this book

This book is more like a mystery with secrets in every chapter. Don't miss out on your chance to explore them.

Mark Lyons works for Dassault Systèmes as the Product Manager for DraftSight. He is the principal of The Lyons' Share™ resource center for DraftSight. Prior to this position, he has been the Training Specialist for DraftSight since December 2008. Before taking this position, Mark was employed as CAD instructor at two regional high schools in Massachusetts. He began his career as a Mechanical Draftsman creating drawings the old fashioned way, with pencil and paper. He has worked in this industry since 1981, and has used many CAD products.

Neb Radojkovic comes from a technical family. Neb started technical (manual) drafting straight after high school in 1979. Over the years and while using his other skills, such as model making and painting, in 1990, Neb was introduced to AutoCAD. From that point on, Neb used AutoCAD to the best of his ability (self-taught) until he finished a 6-months course in AutoCAD 2D and 3D in 2000. His first job as an AutoCAD Operator was in the Doors and Windows manufacturing firm. Soon enough Neb was in search of more challenge, so he became a sole AutoCAD Draftsperson in a Kitchen Manufacturing business. He became expert in cabinetry, millwork and Interiors Drafting and Design using several software: AutoCAD, SolidWorks, DraftSight, and ARES Commander, to mention a few. Since 2008 Neb has been the owner and principal of Cadesigneb.com—his Interiors AutoCAD Drafting business. Neb is now also exploring CAD and 3D Photo Realistic Rendering. As a form of relaxation, Neb enjoys making Logos and Website Designs.

The main companies that Neb worked at were:

- Energoprojekt: Belgrade, Serbia
- Elmwood Kitchens: St. Catharines, ON
- Benson Industries: Saanichton, BC
- Cutting Edge Woodworks: Saanichton, BC

www.PacktPub.com

Support files, eBooks, discount offers and more

You might want to visit www.PacktPub.com for support files and downloads related to your book.

Did you know that Packt offers eBook versions of every book published, with PDF and ePub files available? You can upgrade to the eBook version at www.PacktPub.com and as a print book customer, you are entitled to a discount on the eBook copy. Get in touch with us at service@packtpub.com for more details.

At www.PacktPub.com, you can also read a collection of free technical articles, sign up for a range of free newsletters and receive exclusive discounts and offers on Packt books and eBooks.

 **PACKTLiB**®

http://PacktLib.PacktPub.com

Do you need instant solutions to your IT questions? PacktLib is Packt's online digital book library. Here, you can access, read and search across Packt's entire library of books.

Why Subscribe?

- Fully searchable across every book published by Packt
- Copy and paste, print and bookmark content
- On demand and accessible via web browser

Free Access for Packt account holders

If you have an account with Packt at www.PacktPub.com, you can use this to access PacktLib today and view nine entirely free books. Simply use your login credentials for immediate access.

Table of Contents

Preface

DraftSight is a free CAD program that uses the DWG file format. It includes all the main tools to produce precise technical drawings and runs on a wide variety of operating systems, including Windows, Mac and Linux. The DWG file format is the most used CAD (Computer Aided Design) format, widely spread in all areas of technical drawings.

DraftSight runs in 15 languages, including English, French, German, Spanish, Italian, Portuguese, Chinese (traditional and simplified), and Russian. DraftSight is free for any personal or commercial application.

With a practical project and many illustrations, this book addresses all the main phases for executing a mechanical project, including setting up a drawing, drawing and editing with precision, organizing with layers and obtaining valid information from the drawing, applying patterns and dimension, and defining and printing sheets.

What this book covers

Chapter 1, *Introduction to DraftSight*, includes an introduction to DraftSight, where to download it, how to start and configure it, creating and configuring a drawing, and all the commands to open, save and close drawings.

Chapter 2, *Drawing with Precision*, includes all the important tools that allow drawing with precision, namely coordinates and graphical auxiliary tools, how to draw lines and erase entities, and also the most important visualization commands.

Chapter 3, *Starting to Create Projects*, includes the most important commands to start creating projects, including different types of entities creation such as circles, arcs, rectangles, and polygons. This chapter also covers modification, such as moving, rotating, scaling, copying, mirroring, creating parallel entities, and moving vertices.

Chapter 4, Structuring Projects and Following Standards, includes the main entity properties essential to a correct structure of any drawing, layer being the most important. Three other important properties normally controlled by layer are color, linestyles, and lineweights.

Chapter 5, Inquiring Projects and Modifying Properties, includes several commands for obtaining information in the drawing, two commands to modify entities properties, and a command to select entities based on properties values.

Chapter 6, Creating Complex Projects, includes several commands that allow specific functions important to create complex projects, such as text and text styles, polylines, equally spaced copies, ellipses, rings, revision clouds, tables and table styles, joining and splitting entities.

Chapter 7, Creating and Applying Components, includes commands related to the creation, insertion and other operations about components (blocks). A component, or block, is a set of entities belonging to one or more layers that are grouped together and that can be used as a single element.

Chapter 8, Applying Fills and Patterns, presents hatches, which represent sections, cuts, or materials. Hatches can be a regular pattern composed by families of lines, a single color, or a gradient between two colors.

Chapter 9, Documenting Projects, includes the commands used to document or dimension a drawing, such as linear dimensions, angular dimensions, radius and diameters, ordinate and leader dimensions. Also included are the commands to edit dimensions and dimension styles.

Chapter 10, Printing Efficiently, includes the preparation of sheets for printing, previewing and printing drawings. The concept of sheet, and it's advantages, is introduced.

Chapter 11, Advanced Tools, introduces some advanced concepts and commands, namely referencing other drawings and images, and additional layer commands, very useful when dealing with complex drawings.

What you need for this book

To correctly follow this book and realize all exercises, we need to have DraftSight software, preferably the latest version (V1R3.1 or later). Readers must also download Exercise files from the book's webpage.

Who this book is for

This book is intended for everyone who wants to create accurate 2D drawings in a DWG format file. Examples are from the Mechanical area, but the book can be also useful to architectural, engineering or design professionals, and students. Only some basic computer knowledge, such as dealing with files or using a mouse, is required.

Conventions

In this book, you will find a number of styles of text that distinguish between different kinds of information. Here are some examples of these styles, and an explanation of their meaning.

Code words in text are shown as follows: "To close the current drawing without closing DraftSight, the CLOSE command should be applied."

Any command-line input or output is written as follows:

```
: DELETE
Specify entities» Selection
6 found, 6 total
Specify entities» Enter
```

New terms and **important words** are shown in bold. Words that you see on the screen, in menus or dialog boxes for example, appear in the text like this: "To control the multiple angle, the easiest method is pressing the mouse right button over the **Polar** button and selecting **Settings**".

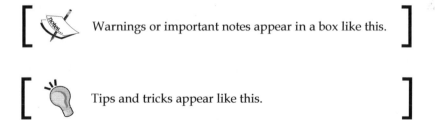

Warnings or important notes appear in a box like this.

Tips and tricks appear like this.

Reader feedback

Feedback from our readers is always welcome. Let us know what you think about this book—what you liked or may have disliked. Reader feedback is important for us to develop titles that you really get the most out of.

To send us general feedback, simply send an e-mail to feedback@packtpub.com, and mention the book title via the subject of your message.

If there is a topic that you have expertise in and you are interested in either writing or contributing to a book, see our author guide on www.packtpub.com/authors.

Customer support

Now that you are the proud owner of a Packt book, we have a number of things to help you to get the most from your purchase.

Downloading the example code and graphics

You can download the example code files for all Packt books and colored graphics of this book you have purchased from your account at http://www.packtpub.com. If you purchased this book elsewhere, you can visit http://www.packtpub.com/support and register to have the files e-mailed directly to you.

Errata

Although we have taken every care to ensure the accuracy of our content, mistakes do happen. If you find a mistake in one of our books—maybe a mistake in the text or the code—we would be grateful if you would report this to us. By doing so, you can save other readers from frustration and help us improve subsequent versions of this book. If you find any errata, please report them by visiting http://www.packtpub.com/submit-errata, selecting your book, clicking on the **errata submission form** link, and entering the details of your errata. Once your errata are verified, your submission will be accepted and the errata will be uploaded on our website, or added to any list of existing errata, under the Errata section of that title. Any existing errata can be viewed by selecting your title from http://www.packtpub.com/support.

Piracy

Piracy of copyright material on the Internet is an ongoing problem across all media. At Packt, we take the protection of our copyright and licenses very seriously. If you come across any illegal copies of our works, in any form, on the Internet, please provide us with the location address or website name immediately so that we can pursue a remedy.

Please contact us at copyright@packtpub.com with a link to the suspected pirated material.

We appreciate your help in protecting our authors, and our ability to bring you valuable content.

Questions

You can contact us at questions@packtpub.com if you are having a problem with any aspect of the book, and we will do our best to address it.

1

Introduction to DraftSight

DraftSight is a free CAD program that uses the DWG file format. It includes all the main tools to produce precise technical drawings and runs on a wide variety of operating systems, including Windows, Mac, and Linux.

We will cover the following topics in this chapter:

- Downloading and starting DraftSight
- Understanding the user interface and locating all elements
- Configuring DraftSight
- Setting up a drawing
- Opening, saving, and closing drawings

Downloading and starting DraftSight

The DWG file format is the most used **Computer Aided Design (CAD)** format. It is used in all areas of technical drawings. Another drawing file format is DXF, mainly used for communication between CAD programs. There are several programs that work with DWG and DXF files, with similar commands and work processes, with DraftSight being among them.

Dassault Systèmes (www.3ds.com), one of the two major CAD companies in the world, licenses the ARES Commander software from Graebert GmbH (www.graebert.com) and uses it as the CAD engine. The first version was released in February 2011. Currently there are versions for Windows 32-bit, Windows 64-bit (XP, Vista, and 7), Mac OS X, and several Linux operating systems (Fedora, Suze, Mandriva, or Ubuntu).

Actually, there are 15 languages available, including English, French, German, Spanish, Italian, Portuguese, Chinese (traditional and simplified), and Russian. DraftSight is free for any personal or commercial application.

DraftSight can be downloaded from www.draftsight.com, selecting the proper operating system. The executable file is around 100 MB and installs very fast. It does not prompt for the language selection, thus installing the local language, but this can be changed after starting DraftSight.

After installation, DraftSight can be initiated by double-clicking the DraftSight desktop icon or by accessing the Windows Start menu and navigating to **All Programs | DassaultSystemes | DraftSight**.

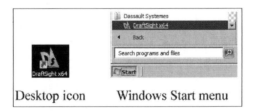

Desktop icon   Windows Start menu

Downloading the example code and graphics

You can download the example code files for all Packt books and colored graphics of this book you have purchased from your account at http://www.packtpub.com. If you purchased this book elsewhere, you can visit http://www.packtpub.com/support and register to have the files e-mailed directly to you.

Understanding the user interface

When starting DraftSight, its graphical user interface is displayed with a blank new drawing called NONAME_0.DWG. The interface can be customized; it contains the following elements by default, which are displayed in the screenshot in the following section:

- **Drawing area**: This area contains the drawing, where graphical entities are inserted, viewed, modified, inquired, or deleted. All colors can be configured.

- **Cursor**: The cursor is controlled by the mouse or other pointing device, which will allow specifying points to create, modify, or select entities, also to access menus or toolbars. The cursor coordinates are displayed at the status line.

- **Coordinate system icon**: This icon displays the positive directions for the X and Y axis. DraftSight, as with other CAD programs, uses a Cartesian coordinate system, where each point is perfectly identified by a pair of coordinates.

- **Main menu**: This top pull-down menu includes most DraftSight commands, divided by categories.

- **Standard toolbar**: This toolbar includes the main commands for starting, opening, saving or printing drawings, for cutting, copying, or pasting entities, for painting properties between entities, undoing and redoing, for visualization operations, and controlling the properties palette.

- **Layers toolbar**: This toolbar includes the LAYER command and the layers list.

- **Properties toolbar**: This toolbar includes the default color, default line type, and default lineweight for new entities or for selected entities. It is advisable to maintain these **ByLayer** (controlled by layer, thus not explicitly).

- **Draw toolbar**: This toolbar includes the main drafting commands, such as creating lines, circles, arcs, and text.

- **Modify toolbar**: This toolbar includes the main modifying commands, such as deleting, moving, copying, and rotating.

- **Properties palette**: Some DraftSight commands display a palette, which can always be visible. This palette can be turned off, as it will not be required now.

- **Model/Sheets tabs**: Drawings are made on the model space. Sheets spaces are specific to prepare sheets for printing; presented in *Chapter 10, Printing Efficiently*.

- **Command window**: This is one of the most important interface areas. It includes all requests from the program, displaying available options, and where users can write command names, their aliases, or options. The last line indicates that DraftSight is prompting the colon (:), which means it is waiting for a command. The command window can be resized or moved.

- **Status bar**: This bar, beneath the command window, includes the cursor coordinates and the access to drafting auxiliary tools, presented in the next chapter. When hovering over a toolbar icon or a menu item, a small help and the command name are displayed on the left.

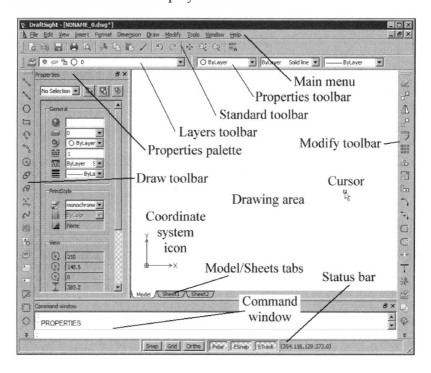

Configuring DraftSight

First, we may choose a different language for using DraftSight. When installing, a language based on user's location is automatically applied, which may not be the desired one. To choose a different language, the _LANGUAGE command must be applied (without forgetting underscore before command name) along with the language corresponding number (a question mark displays all available). Then, reinitiating DraftSight will set up the chosen language.

It is very simple to configure DraftSight. The OPTIONS command includes almost all settings and system options. We can access the command by digitizing its name, OP as its alias, choosing **Options** on the menu displayed by pressing the mouse right button, or **Options** on the **Tools** main menu.

The command displays a dialog box with several sections on the left column. Choosing one of these sections, the corresponding options are displayed on the right area with a tree structure:

- **File Locations**: This section allows defining the location of important files and specific file names, including the **Support Files Search Path**, default **Drawing Files Location**, or **External References Files Location**.

- **System Options**: This includes general options related to the system, such as Element **Colors** to configure interface colors (displayed in the following screenshot), DWG version for saving, **Printing** options, or **Auto-save & Backup** options.

- **User Preferences**: Here we can define several **Drafting** and **Mouse** user options, as well as aliases for commands.

- **Drawing Settings**: This includes settings specific to the current drawing. The most important settings are presented in the following screenshot.

- **Drafting Styles**: This section includes **Active Drafting Styles** and the possibility to create or modify text styles, dimension styles, rich line (multiline) styles, and table styles.

- **Profiles**: Here it is possible to manage user profiles in order to easily change multiple options at once.

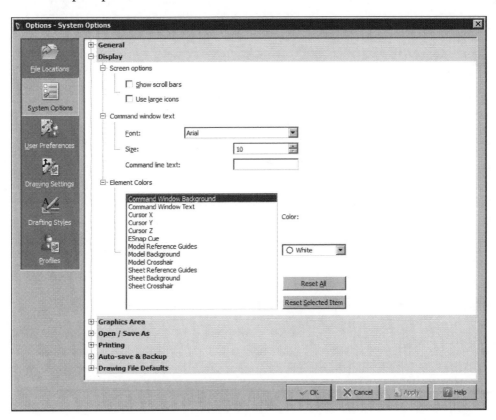

The default drawing area is black. A white drawing area was adopted throughout this book to improve images visualization.

Starting and setting up a drawing

After configuring DraftSight, let's start a new drawing and configure it.

Starting a drawing

The NEW command (shortcut *Ctrl + N*, on the **Standard** toolbar, or **File** main menu) allows us to start a drawing. The command displays a standard file dialog box in order to choose the template.

When starting a new drawing, we may start from a blank drawing (any one coming with DraftSight) or from a template with some configurations already done. Templates are drawing files with the DWT extension by default, placed at C:\Users\user\AppData\Roaming\DraftSight\1.2.265\Template or similar, depending on the operating system and DraftSight version. With the OPTIONS command, going to the **File Locations** | **Drawing Support** | **Drawing Template File Location**, we may change this location.

To create a template, we prepare the drawing to be used and then apply the SAVEAS command by selecting **Drawing Template (*.dwt)** on **Save as type** list.

The SMARTNEW command (no alias or icon) allows us to start a new drawing without displaying the file dialog box. We may configure which template will be applied with the OPTIONS command, **System Options** | **Open / Save As** | **Template file name for SmartNew**.

Setting up a drawing

When using a blank template (for instance, standardiso.dwt, coming with DraftSight), we may configure it, especially units. Again with the OPTIONS command, going to **Drawing Settings** | **Unit System**, we can configure the following drawing options:

- **Base angle**: We control the base angle direction, from which absolute angles are measured, by clicking on the compass or by writing a value. By default, 0° is East (3 o'clock). Normally, angles are measured positively in the counter clockwise direction, unless **Clockwise** is checked.

- **Length**: The **Type** list allows the linear unit to be defined and the **Precision** resolution (number of decimal places or fractions) to be displayed in coordinates or answer to inquiry commands.

- **Angle**: The **Type** list allows the angular unit to be defined and the **Precision** resolution (number of decimal places or other) to be displayed in coordinates or answer to inquiry commands.

- **Units scale**: The **Block units format list** identifies the drawing unit when inserting contents like blocks or images. With this option, DraftSight automatically applies the suitable scale factor when inserting a block whose unit is different from the drawing unit. A **block** is a set of objects constituting a single object, like a door or a bolt; its creation or modification is presented in *Chapter 7, Creating and Applying Components*.

- **Preview**: This simply displays a preview of linear and angular units.

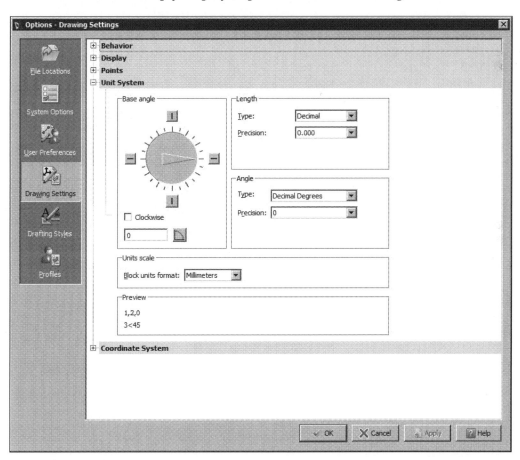

By digitizing the UNITSYSTEM command name we get access directly to this section of the OPTIONS command. But a better way is to simply digitize UN (or UNITS).

Opening, saving, and closing drawings

The OPEN command (shortcut *Ctrl + O*, on the **Standard** toolbar or **File** main menu) opens a drawing in formats DWG or DXF. Also, it allows opening template files, DWT extension, to edit these.

It only displays a standard file dialog box to select the file, including a preview area.

DraftSight allows several open drawings. To activate another open drawing, we may press *Ctrl + Tab* or apply the **Window** main menu and select the drawing by its name.

When saving drawings, there are two commands depending on whether or not you want to modify the name or location.

The SAVE command (shortcut *Ctrl + S*, on the **Standard** toolbar, or **File** main menu) saves the current drawing, maintaining its format, name and location. If the current drawing has no name, the next command is automatically applied.

If we want to modify the name, location, type of file, or version, the SAVEAS command (shortcut *Ctrl + Shift + S* or the **File** main menu) must be used. This command displays a standard file dialog box and we can specify name, location and type/version of drawing file. On the **Save as type** list, as displayed on next image, there are several possibilities covering all DWG and DXF versions for last twenty years. Saving as a template, DWT file is also available.

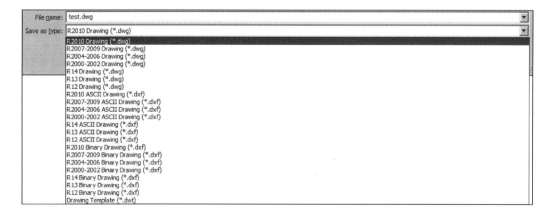

There is a third command to save drawings, called SAVEALL (no alias and not on menus). This one allows us to save all open drawings at once.

To close the current drawing without closing DraftSight, the CLOSE command (**File** main menu)should be applied. This is equivalent of pressing the lower **X** button on the upper-right corner of the user interface. If the drawing has modifications that have not yet been saved, a warning box is displayed, reminding us to save the drawing.

The CLOSEALL command (**Window** main menu) allows us to close all drawings without closing DraftSight. To each drawing that has modifications not yet saved, a warning box is displayed, allowing us to save that drawing.

To quit DraftSight, the EXIT command should be used. This is the equivalent of pressing the upper **X** button on the upper-right corner of the user interface. If any drawing has modifications that have not yet been saved, a warning box is displayed, allowing us to save that drawing.

Summary

In this chapter, we covered an introduction to the free DraftSight program, including where to download it from, how to start and configure it, followed by creating and configuring a drawing. Then all commands to open, save, and close were presented.

In the next chapter we will learn how to draw lines with precision, delete entities, and how to apply pans and zooms to better view drawings.

2
Drawing with Precision

We are going to start drawing with precision, drawing lines, and learning how to delete entities. Depending on project dimensions, it may be necessary to zoom in or out, also explained in this chapter.

In this chapter we will cover the following topics:

- Defining coordinates
- Drawing lines
- Deleting lines and other entities
- Applying auxiliary tools to draw with precision
- Visualizing the drawing

Defining coordinates

An essential step in all CAD programs is to draw with precision. Defining entity points like endpoints of a segment line or a circle center, without any doubt is a constant process, which can be achieved by introducing coordinates, absolute or relative. There are also several auxiliary tools that ease the process, as shown later in the chapter.

Absolute coordinates

DraftSight uses a Cartesian system, with an origin and three orthographic axes: X, Y, and Z. All points are referenced with coordinates related to these axes. In 2D drawings, the Z coordinate can be omitted.

So, to define a 2D point in absolute coordinates, two values must be set: the first along the X axis and the second along the Y axis, separated by a comma:

`X_value,Y_value`

Examples, as displayed in following diagram, are the points `-2,3`; `-3,-2`; `1,-2`; or `4,2`.

 Absolute coordinates are mainly used with georeferenced drawings or in 3D models. In Mechanical Drafting normally there is no need for these coordinates.

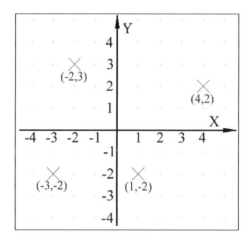

Relative Cartesian coordinates

Often, points are defined relating to the last introduced point. To specify a point in relative Cartesian coordinates, an *at the rate* symbol (@) must be introduced, followed by increment in X direction, a comma, and increment in Y direction:

`@increment_X,increment_Y`

Of course one of these values may be zero. If both are zero (`@0,0`), this means that the point is coincident with the last one. For instance, in the next image, if point A is the last point defined, point B can be given by `@2,1` and point C, after B being introduced, can be given by `@-1,2`.

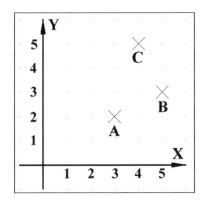

Relative polar coordinates

Sometimes, increments between points are not known. Instead we know the direct distance and angle between points. In these situations, it is preferable to apply relative polar coordinates, starting by the @ symbol, followed by the distance, left angle bracket as the angle symbol, and the absolute angle:

`@distance<angle`

Angles, as displayed in following diagram, by default are in decimal degrees, measured positive counter-clockwise and the zero degrees corresponds to East.

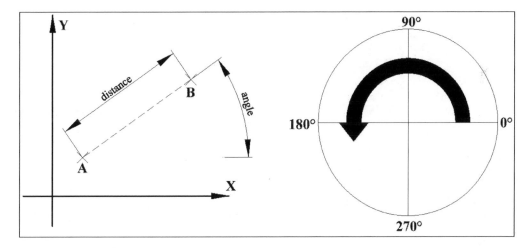

 By applying auxiliary tools **Ortho, Polar,** or **ETrack**, points specification is simplified, reducing the number of points that should be defined by relative coordinates.

Drawing lines and deleting

Let's start drawing lines with precision. After trying this command, several lines should be deleted, thus introducing entities selection.

Drawing lines

The LINE command (alias L, 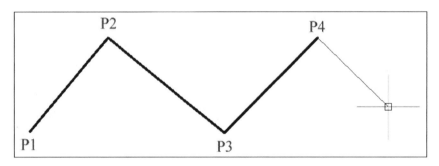 on the **Draw** toolbar, or **Draw** on the main menu) creates segment lines. By default, it prompts only for points and connect them with separate lines. When done, press *Esc* or *Enter* to end the command:

```
LINE
Options: Segments or
Specify start point» P1
Options: Segments, Undo, Enter to exit or
Specify next point» P2
Options: Segments, Undo, Enter to exit or
Specify next point» P3
Options: Segments, Undo, Close, Enter to exit or
Specify next point» P4
Options: Segments, Undo, Close, Enter to exit or
Specify next point» Esc
```

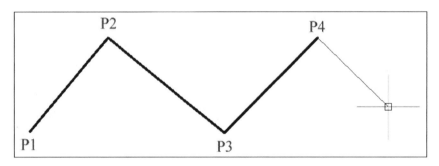

Points can be defined by pressing the left mouse button or by coordinates. Later in this chapter we will make a drawing with coordinates.

Without ending the command, the Undo option allows for undoing the last defined point. It can be applied several times until undoing the first point in the command. After defining three points, the Close option may be used to close the polygon. Upon entering the command, if *Enter* is pressed, the line starts automatically at the last point of the last line, arc or polyline drawn.

Selecting and deleting

Many DraftSight commands prompt for selecting entities. For instance, if entities are to be moved, rotated or deleted, the first thing to do is select these. There are several options to select entities, but for now let's see the three most used. Other useful selection options are presented in next chapter. Entities, while selected, are represented with a temporary dashed line type.

- **Picking on top of an entity**: Entities are selected one by one by picking on top of each one.

- **Window**: By opening a rectangle to the right (by default, it assumes a blue color), all entities that remain completely within are selected.

- **Crossing**: By opening a rectangle to the left (by default, it assumes a green color), all entities that remain completely within or crosses the rectangle are selected.

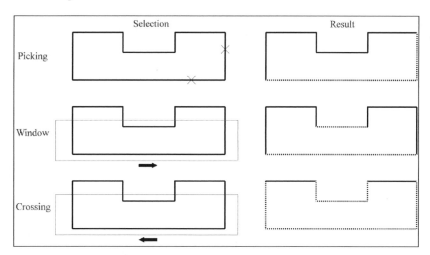

To remove entities from selection, the easiest way is to press the *Shift* key and select it by picking window or crossing.

By default, DraftSight accepts entities selection before specifying a command. If an entity is picked without command, it is immediately selected. When picking on a blank area, a rectangle selection is open: to the right performs a window selection, to the left performs a crossing selection. When selecting entities before a command, small blue squares are displayed on selected entities, called grips, whose utility will be presented later.

The DELETE command (alias DEL, *Delete* key on the **Modify** toolbar or **Modify** main menu) deletes the selected entities. It only prompts for the selection and, for each operation, indicates the number of selected objects and total number. When done, press *Enter* to end the command:

```
: DELETE
Specify entities» Selection
6 found, 6 total
Specify entities» Enter
```

 This command also accepts ERASE and the alias E, as with other similar programs.

Graphical auxiliary tools

DraftSight includes some graphical auxiliary tools, presented next, that considerably case the process of defining points with precision. These tools are accessed by buttons located on the status bar, at the bottom of the graphical interface or by the keyboard function keys.

 The auxiliary tools settings can be modified in the middle of a command.

Ortho and Polar

When Ortho function (*F8* key) is on, the cursor locks on the X or Y directions (normally horizontal and vertical) and the obtained points related to the previous point are orthogonally aligned.

To define points that are horizontally or vertically aligned to previous points with Ortho On, we just direct the cursor to the wanted direction and type the distance. For instance, instead of defining a point @10, 0, we turn on Ortho, move the cursor to the right, type 10 and press *Enter* or the Space bar.

The `Polar` function (*F10* key) is similar, but allows for locking directions with angles which are not multiples of 90 degrees. For instance, an incremental angle of 15 degrees may be useful for some parts of a project. To control the multiple angle, the easiest method is to press the right mouse button over the **Polar** button and to select **Settings**. It opens the OPTIONS dialog box, **User Preferences** area, **Drafting Options**, **Polar Guides**. The **Incremental angles for Polar guide display** list allows the user to select the incremental angle. It is also possible to add other angles, not incremental, by pressing the **Add** button:

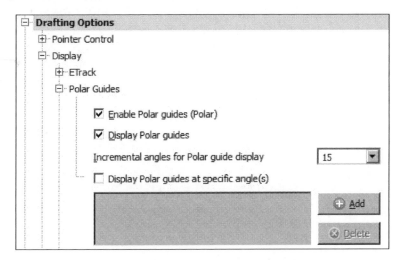

The following diagram displays lines drawn with the help of the `Ortho` and `Polar` functions.

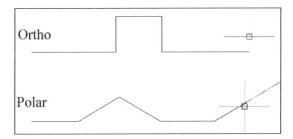

Entity snaps

Entity snaps are the characteristic points of entities, such as endpoints, centers, tangents, and so on. When DraftSight prompts for a point, it is possible to access entity snaps with all precision. When pausing over an entity snap, a small label is displayed, identifying the snap. These points may be accessed automatically or manually.

- **Automatic access (default entity snaps)**: If ESnap function (*F3* key) is on and depending on the default snaps, whenever DraftSight prompts for a point, if the cursor moves over an entity, it may snap to a point on that entity. To control which entity snaps are by default, the easiest process is, like the Polar function, to press the right mouse button over the ESnap button and to choose **Settings**. It opens the OPTIONS dialog box, **User Preferences** area, **Drafting Options**, **Pointer Control**. The same box can be opened with the ENTITYSNAP command (alias ES or OS).

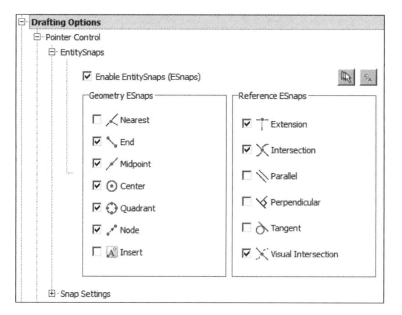

- **Manual access**: We should only access by default the most common entity snaps. Activating all of them would be a complete mess. The less frequently used entity snaps should only be accessed manually. Whenever DraftSight prompts for a point, by pressing *Ctrl* and the right mouse button, the manual entity snaps menu appears and we may select an entity snap for only the next point. In this case, the automatic access is suspended. Besides this menu, manual entity snaps can be accessed by digitizing the first three letters (for instance, PER for **Perpendicular**) or the **Entity Snap** toolbar.

The following diagram includes examples of the most common entity snaps:

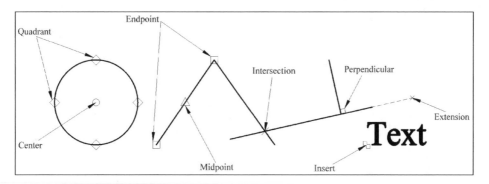

 The **Extension** entity snap works differently from the others, as we must pause over an endpoint, move the cursor in the entity direction and then write a distance or pick a point.

The manual entity snaps menu include two interesting options to obtain points: **From**, to obtain a point with relative coordinates from a base point and **Mid Between 2 Points**, to obtain a point midway from two points.

Entity snap tracking

Another very useful function is `ETrack` (*F11* key). This one allows us to derive a point from default entity snaps. This means that `ESnap` function (*F3* key) must be on. We pause over a default entity snap, acquiring that point, and orient the cursor in any direction of the `Polar` function. A provisory dashed line appears. Then a distance along that line can be given.

It is possible to acquire two points and obtain intersections between two provisory lines.

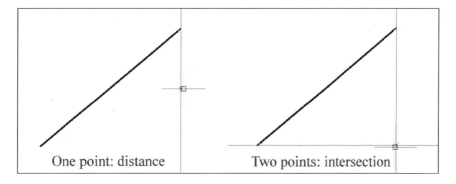

Snap and Grid

The two remaining functions, when both are on, allow for quick schematic drawings.

The Grid function (*F7* key) displays a dot grid in the drawing area. Knowing its spacing, we get an idea for general dimensions.

The Snap function (*F9* key) allows for discrete cursor movement. If having the same Grid spacing, this means that the cursor only jumps between Grid points.

To modify spacing, the easiest method is by pressing the right mouse button over the Snap or Grid buttons and selecting **Settings**. It opens the OPTIONS dialog box, **Drafting Options**, with the relevant options displayed.

Visualization tools

In CAD, it is advisable to always draw at 1:1 scale. In order to get the desirable precision, zooming and panning while drawing or consulting is a continuous task. Having a wheel mouse or other advanced input device with scroll functionality, zooming and panning are a straightforward process. With a wheel mouse we can:

- **Zoom in/zoom out**: Rotating the wheel forward we zoom in, we see more detail. Rotating the wheel backward we zoom out, we see less detail and more of our model.
- **Zoom fit**: With a quick double-click on the wheel, we see the drawing extension. This means the area occupied by entities in the model.
- **Pan**: By holding down the wheel and dragging the mouse, we slide the view, maintaining the same zoom factor.

Zooming and panning can be used transparently, in the middle of a command.

The ZOOM command (alias Z, **Zoom** toolbar or **View** main menu) allows us to modify the visualization scale without a wheel mouse, thereby presenting several options. By default, it prompts for defining a rectangular area, zooming in to that area:

```
: ZOOM
Default: Dynamic
Options: Bounds, Center, Dynamic, Fit, Left, Previous, SElected, specify
a scale factor (nX or nXP) or
Specify first corner» Point
Specify opposite corner» Point
```

Instead of picking a point, if pressing *Enter*, we enter in mode **Dynamic**, where dragging (maintaining pressure on the left mouse button) upwards is equivalent to zoom in, and dragging downwards is equivalent to zoom out. Other useful options:

- **Fit**: Displays all entities in the drawing, that is, all drawing is presented with maximum scale.

- **Previous**: Displays the previous visualization. It is possible to come back to the last ten visualizations.

DraftSight includes some commands that act directly: ZOOM command and a specific option. For instance, ZOOMBACK (alias ZB) gets the previous view; ZOOMFIT (alias ZF) zooms to fit.

The PAN command (alias P, on the main toolbar or **View** main menu) allows us to slide the view without modifying visualization scale, by dragging the cursor (maintaining pressure on the left mouse button). We press *Esc* or *Enter* to exit.

With DraftSight, unlike other CAD programs, there is no need for regenerations. However, there is the REBUILD command (alias RE) that recalculates what is displayed on the drawing area.

Exercise 2.1

Only with the LINE command and auxiliary tools, we are going to make the drawing displayed in the following diagram (first angle projection, also known as European projection).

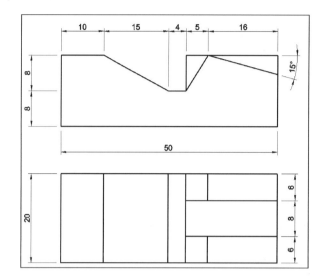

1. With the NEW command, start a new drawing from the standardiso.dwt template.

2. As there is a line at 15 degrees, with the cursor over the Polar button, select **Settings**, apply a 15 degrees **Incremental angles for Polar guide display** and confirm with clicking on **OK**.

3. Turn on Polar, ESnap, and ETrack. We are going to start drawing the front view (above) from the lower-right corner. With the LINE command, click on a first point anywhere, direct the cursor left, and digitize 50. Now direct the cursor up and digitize 16. Direct the cursor right and digitize 10. The first three segments should be drawn.

4. If the LINE command has been cancelled, apply it again (if DraftSight is prompting for a command, the *Enter* key repeats the last used command) and press *Enter* again to start from the last point. Then, applying relative coordinates, digitize @15,-8 to get the next point. Direct the cursor right and digitize 4; again with relative coordinates, @5,8; now direct right and below until the -15 degrees temporary line appears. As we don't have the real distance, pause over the lower-right corner and move the cursor up until getting two intersecting temporary lines, as displayed in next image and click; now click on the lower-right corner. If the LINE command was not interrupted, the **Close** option may be used.

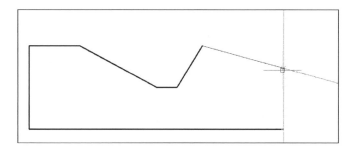

 If any point is not correctly defined, there is no need to exit the
LINE command. Instead, the **Undo** option should be applied.

5. We can now finish the front view. With the LINE command, click on the
    upper-right corner, move the cursor up, pause over the vertex where the 15
    degrees line started and click when getting two crossing temporary lines.
    Pause over the second endpoint of the horizontal line with length 4 and click
    when getting the intersection horizontal. Finally click on the last point where
    we paused.

6. We are now going to create the top view (below). This view must be
    vertically aligned to the front view, so the ETrack function is very useful.
    With the LINE command, pause over the lower-left corner of the front view.
    Move the cursor down and, over the temporary line, pick a point. Direct the
    cursor right and digitize 50. Direct the cursor down and digitize 20. Direct
    the cursor left and digitize 50. To complete the rectangle digitize c to close it.

7. To project edges from the front view, pause over the respective vertex and
    over other corners to get intersections, as displayed on next image.

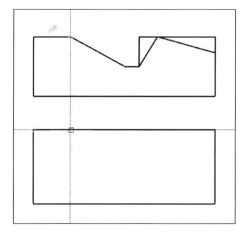

8. Applying the LINE command with auxiliary tools, all three vertical lines on the top view are drawn.

9. To draw the first inner horizontal line, pause over the upper-left corner, direct down and digitize 6. Direct left and get the intersection with the first vertical line.

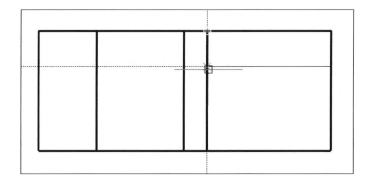

10. We repeat the LINE command to draw the remaining lines.

11. We save the drawing in a proper folder with the name DR02-01.DWG.

Summary

This chapter includes all the important tools that allow drawing with precision, starting with the three types of coordinates used to define points: absolute, relative Cartesian and relative polar.

In order to start drawing, the LINE and DELETE commands were presented, as well as the most common entities selection method. The graphical auxiliary tools include the Ortho function, to specify points orthogonally; the Polar function, to define points over a default multiple angle; entity snaps to get precise entity points; ETrack function to derive points from default entity snaps and over Polar; Snap and Grid functions to draw quick schematic drawings with equally spaced points.

Finally, the most important visualization commands were introduced, namely ZOOM and PAN, as well as modifying visualization with a wheel mouse.

In the next chapter, the most important creation and modification commands will be presented, allowing us to start creating projects.

3

Starting to Create Projects

After covering the LINE and DELETE commands presented in the previous chapter, we are now adding several commands essential to create and modify our projects.

In this chapter we will cover the following topics:

- Drawing curved entities, namely circles and arcs
- Drawing rectangles and regular closed polygons
- Additional methods for selecting entities
- Moving and copying entities
- Applying symmetries
- Rotating and scaling
- Creating parallel entities
- Stretching and shortening entities
- Trimming and extending entities

Drawing circles and arcs

The CIRCLE command (alias C, ⊙ on the **Draw** toolbar, or **Draw** main menu) creates circles. By default, it prompts for the circle's center point and radius, but other construction options are available. The radius can be specified by value or by point:

```
: CIRCLE
Options: 3Point, 2Point, Ttr, TTT, Enter to exit or
Specify center point» P1
Default: 15
Options: Diameter or
Specify radius» P2
```

After specifying the center point, the Diameter option allows the user to define a circle by center and diameter. If the center location is unknown, there are four construction options: 3Point defines a circle from three not collinear points; 2Point defines a circle by two points on diameter; Ttr defines a circle that is tangent to two existing entities and a specified radius; TTT defines a circle that is tangent to three existing entities.

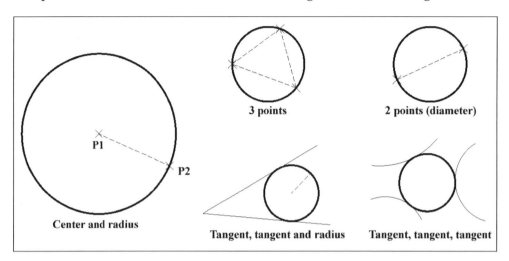

3 points   2 points (diameter)

P1

P2

Center and radius   Tangent, tangent and radius   Tangent, tangent, tangent

The ARC command (alias A, ⟋ on the **Draw** toolbar, or **Draw** main menu) creates arcs. By default, it prompts for three points, **start**, **through** and **end**.

```
: ARC
Options: Center, Append, Enter to continue from last point or
Specify start point» P1
Options: Center, End or
Specify through point» P2
Specify end point» P3
```

There are eleven methods for creating arcs, as can be seen from the **Draw | Arc** main menu. Basically, each method needs three parameters. The remaining parameters are: Center to define the arc center point; Angle to define the included angle; Length to define the arc chord; Radius to define the arc radius; Direction to define the tangent direction on the first point.

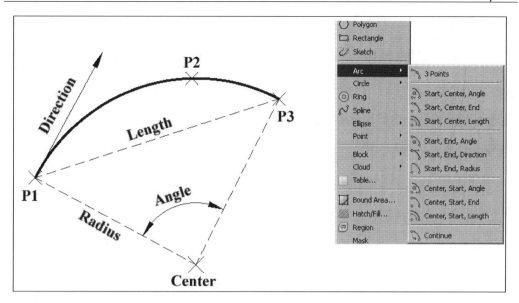

For users starting to learn CAD with DraftSight, it is advisable to apply the command through the main menu. Here, it is easier to select the best method and parameter's order.

Several methods allow for drawing an arc and its complementary arc. In these methods, DraftSight draws the arc in the direct direction (counter clockwise). But when applying methods with `Direction` or `Angle` (with a negative value), it is possible to draw it clockwise.

Drawing rectangles and polygons

The RECTANGLE command (alias REC, ⬜ on the **Draw** toolbar, or **Draw** main menu) creates polylines with a rectangle shape. A polyline, as it will be detailed later, is a single entity composed by a continuous sequence of segments and/or arcs. By default, it prompts for two opposite corners:

```
: RECTANGLE
Options: Chamfer, Elevation, Fillet, Thickness, line Width or
Specify start corner» P1
Options: Area, Dimensions, Rotation or
Specify opposite corner» P2
```

The Chamfer option allows the user to apply a diagonal cut in all four corners. The Fillet option allows the user to round all corners. Elevation and Thickness are only for 3D, thus are not useful. line Width allows defining a width for the rectangle. After defining the first corner, additional options are available: Area allows the user to define a rectangle by one of the sides and area. Dimensions defines the rectangle by horizontal and vertical dimensions. Rotation defines a rotated rectangle.

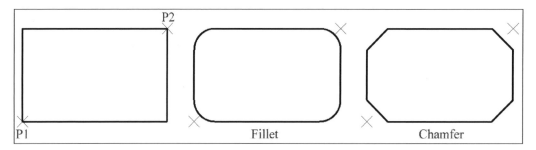

The POLYGON command (alias POL, ⬡ on the **Draw** toolbar, or **Draw** main menu) creates polylines with a regular closed polygonal shape. By default, it prompts for the number of sides, center point (if the distance is to a vertex or to a mid point), and distance:

```
: POLYGON
Default: 4
Specify number of sides» 6 or other number
Options: Side length or
Specify center point» P1
Default: COrner
Options: COrner or Side
Specify distance option» COrner
Specify distance» P2
```

The command has three options, as displayed in next image:

- Side length: This allows the user to create a polygon from two points that define a side
- COrner: This specifies that the distance is from the polygon center to a vertex
- Sides: This specifies that the distance is from the center to a mid point

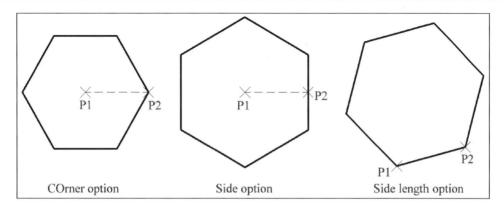

| COrner option | Side option | Side length option |

 When applying the Side length option, the order of points specifies to which side the polygon is created, always in the counter clockwise direction.

Selecting entities

In the previous chapter, we learned three processes to select entities: picking on top, window (selection rectangle to the right), and crossing (selection rectangle to the left). If the selection occurs answering to a command prompt, any selection operation displays how many objects were selected and the total of selected objects:

```
: DELETE
Specify entities»Selection
1 found, 1 total
Specify entities»Selection
2 found, 3 total
```

There are several other selection methods. The following list contains some more that may be useful in some situations:

- **ALL**: This method selects all entities in the drawing, with the exception of entities in layers locked or frozen (layers are presented in next chapter)
- **Last**: This method selects the last entity added to the drawing, either created or by an editing command as COPY, MIRROR or ARRAY
- **Previous**: This method selects the last selection set
- **CRossline/Fence**: This method prompts for points defining temporary lines and selects all entities crossed by these lines

- **CPolygon**: This method selects all entities inside or that cross a polygon defined by a sequence of points.

- **WPolygon**: This method selects all entities completely inside a polygon defined by a sequence of points.

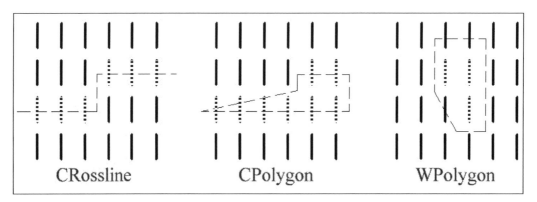

CRossline   CPolygon   WPolygon

Moving, copying, and reflecting

The MOVE command (alias M, 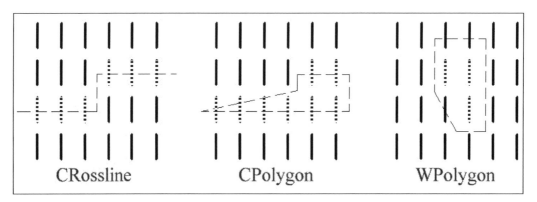 on the **Modify** toolbar, or **Modify** main menu) moves the selected entities. By default, it prompts the entities selection and two points that define the displacement:

```
: MOVE
Specify entities» Selection
2 found, 2 total
Specify entities» Enter
Default: Displacement
Options: Displacement or
Specify from point» P1
Options: Enter to use from point as displacement or
Specify destination» P2
```

The second point can be directly picked with the cursor, by relative coordinates or by applying a distance along the ortho or polar directions.

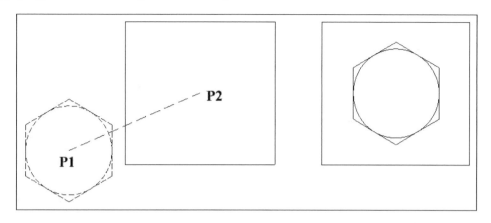

The COPY command (alias CO, 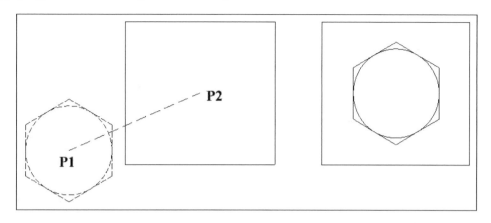 on the **Modify** toolbar, or **Modify** main menu) copies the selected entities. By default, it prompts the entities selection, a base point and one or more destination points, and we press *Enter* to exit:

```
: COPY
Specify entities» Selection
2 found, 2 total
Specify entities» Enter
Default: Displacement
Options: Displacement or
Specify from point» P1
Options: Enter to use first point as displacement or
Specify second point» P2
Default: Exit
Options: Exit, Undo or
Specify second point» Enter
```

Instead of picking the second point with the cursor, it is frequent to apply relative coordinates or a distance along the ortho or polar directions.

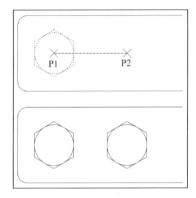

The MIRROR command (alias MI, 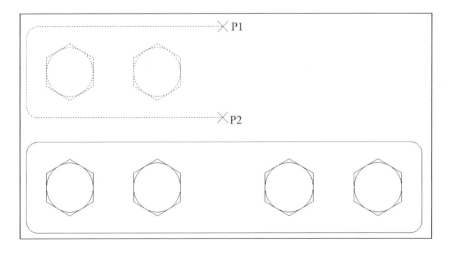 on the **Modify** toolbar, or **Modify** main menu) applies a reflection to the selected entities. By default, it prompts the entities selection, two points that define the mirror line and if original entities are deleted:

```
: MIRROR
Specify entities» Selection
5 found, 5 total
Specify entities» Enter
Specify start point of mirror line»P1
Specify end point of mirror line»P2
Default: No
Confirm: Delete source entities?
Specify Yes or No»No
```

In almost all technical areas, particularly in mechanical drafting, symmetry situations are frequent. By drawing only half or a quarter of the project, and then applying the MIRROR command, productivity can rise significantly.

Rotating and scaling

The ROTATE command (alias RO, on the **Modify** toolbar, or **Modify** main menu) allows the user to rotate the selected entities around a point. By default, it prompts the entities selection, center of rotation (pivot point) and angle of rotation, positive in counter-clockwise direction:

```
: ROTATE
Active positive angle in CCS: DIRECTION=counterclockwise BASE=0
Specify entities» Selection
2 found, 2 total
Specify entities» Enter
Specify pivot point» P1
Default: 0
Options: Reference or
Specify rotation angle» Angle value
```

The Reference option allows the user to define the rotation angle calculated from an existing angle (normally by two points) and a final angle (normally typed in).

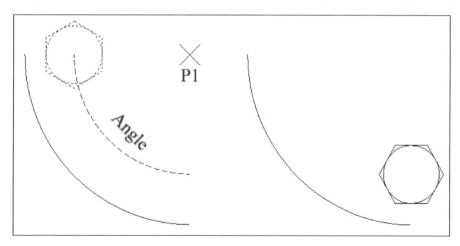

Unlike other DWG programs, in DraftSight this command has no Copy option. Neither does it allow the user to define the final angle, in Reference, by two points.

The SCALE command (alias SC, on the **Modify** toolbar, or **Modify** main menu) allows the user to apply a scale factor to the selected entities related to a base point. By default, it prompts the entities selection, base point and scale factor:

```
: SCALE
Specify entities» Selection
1 found, 2 total
Specify entities» Enter
Specify base point» P1
Options: Reference or
Specify scale factor» Value
```

If scale factor is greater than one, entities will become bigger, if less than one, entities will become smaller. The Reference option allows defining the scale factor calculated from an existing distance (normally by two points) and a final distance (normally typed in).

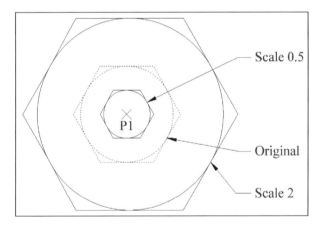

 As with the previous command, this one has no Copy option. Neither does it allow the user to define the final distance, in Reference, by two points.

Offsetting and stretching

The OFFSET command (alias o, 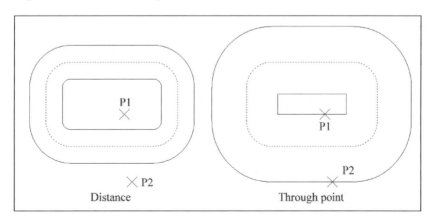 on the **Modify** toolbar, or **Modify** main menu) allows the user to create parallel entities. By default, it prompts for the distance, one entity selection and to which side is the parallel entity. Entity selection and side prompts are repeated until pressing *Enter* or *Esc* keys:

```
: OFFSET
Active settings: Delete source=No Layer=Source
Default: 1
Options: Delete, DIstances, destination Layer, Through point or
Specify distance» Value
Default: Exit
Options: Exit, Undo or
Specify source entity»Selection
Default: Exit
Options: Both sides, Exit, Multiple, Undo or
Specify side for destination»P1
Default: Exit
Options: Exit, Undo or
Specify source entity»Selection
Default: Exit
Options: Both sides, Exit, Multiple, Undo or
Specify side for destination»P2
Default: Exit
Options: Exit, Undo or
Specify source entity»Enter
```

Instead of a distance, the Through point option allows parallel entities passing over specified points, thus enabling different distances in a single command. The Delete option allows the user to delete original entities. The **Destination Layer** option allows the parallel entities to be placed in the current layer, instead of source layers. Layers are presented in next chapter.

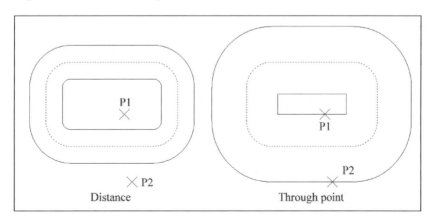

Distance   Through point

The STRETCH command (alias S, on the **Modify** toolbar, or **Modify** main menu) allows the user to move the vertices of entities. Instead of moving the entity, only parts are moved, allowing to increase or decrease entities only in one direction. Vertices to be moved must be selected by Crossing or CPolygon. It prompts for vertices selection and two points specifying the move:

```
: STRETCH
Specify entities to stretch by CWindow or CPolygon...
Specify entities» Selection by crossing
2 found, 2 total
Specify entities» Enter
Options: Displacement or
Specify from point» P1
Options: Enter to use from point as displacement or
Specify destination» P2
```

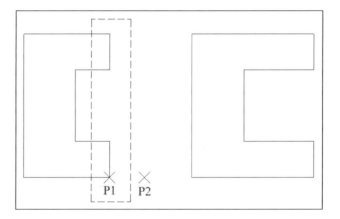

> This is one of the best commands to modify projects, allowing stretching entities in one direction while preserving connections.

Trimming and extending

The TRIM command (alias TR, ✂ on the **Modify** toolbar, or **Modify** main menu) allows the user to cut parts of entities at the edges defined by other entities. By default, it prompts for the selection of cutting edges and, then the entities to cut. *Enter* or *Esc* ends the command:

```
: TRIM
Active settings: Projection=CCS, Edge=None
Specify cutting edges ...
Options: Enter to specify all entities or
Specify cutting edges» P1
1 found, 1 total
Options: Enter to specify all entities or
Specify cutting edges» Enter
Options: Crossing, CRossline, Project, Edge, eRase, Undo, Fence, Shift +
select to extend or
Specify segments to remove» Selection by crossing
Options: Crossing, CRossline, Project, Edge, eRase, Undo, Fence, Shift +
select to extend or
Specify segments to remove» Enter
```

Crossing and CRossline (same as Fence) are selection options, but if we open a selection rectangle to the left, Crossing is directly applied. Project option allows the user to cut entities not lying on the same plane, which is only useful for 3D. Edge option extends virtually cutting edges. eRase allows erasing entities without leaving the TRIM command. Undo undoes last cut.

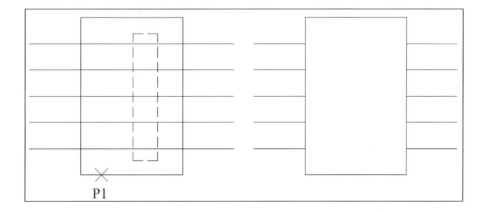

When the command prompts for cutting edges, pressing *Enter* selects all entities in the drawing as potential edges. This is particularly useful when you need to cut entities in different parts of the drawing.

When requesting entities to cut, pressing the *Shift* key and selecting entities, these are extended to edges, thus applying the next command.

The `EXTEND` command (alias `EX`, on the **Modify** toolbar, or **Modify** main menu) allows the user to extend' entities to other entities. By default, it prompts for the selection of boundary edges and then prompts the entities to extend. *Enter* or *Esc* ends the command:

```
: EXTEND
Active settings: Projection=CCS, Edge=None
Specify boundary edges ...
Options: Enter to specify all entities or
Specify boundary edges» P1
1 found, 1 total
Options: Enter to specify all entities or
Specify boundary edges» Enter
Options: Crossing, CRossline, Project, Edge, eRase, Undo, Fence, Shift +
select to trim or
Specify segments to extend» P2
Options: Crossing, CRossline, Project, Edge, eRase, Undo, Fence, Shift +
select to trim or
Specify segments to extend» P3
Options: Crossing, CRossline, Project, Edge, eRase, Undo, Fence, Shift +
select to trim or
Specify segments to extend» Selection by crossing
Options: Crossing, CRossline, Project, Edge, eRase, Undo, Fence, Shift +
select to trim or
Specify segments to extend» Enter
```

`Crossing` and `CRossline` (same as `Fence`) are selection options, but if we open a selection rectangle to the left, `Crossing` is directly applied. The `Project` option allows the user to extend entities not lying on the same plane, which is only useful for 3D. `Edge` option extends virtually boundary edges. `eRase` allows erasing entities without leaving `EXTEND` command. `Undo` undoes the last extend.

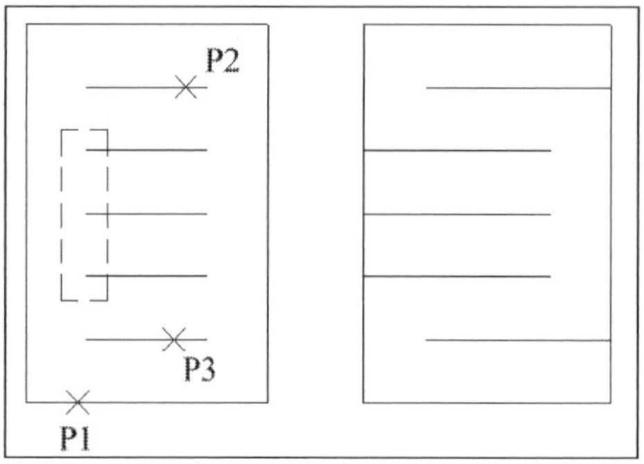

When the command prompts for boundary edges, pressing *Enter* selects all entities in the drawing as potential edges. This is particularly useful when you need to extend entities in different parts of the drawing.

When requesting entities to extend, pressing the *Shift* key and selecting entities, these are cut to edges, thus applying the previous command.

Filleting and chamfering

The FILLET command (alias F, on the **Modify** toolbar, or **Modify** main menu) allows the user to round or obtain acute corners. By default, it prompts only for the selection of two entities and applies the default radius:

```
: FILLET

Mode = TRIM, Radius = 0.5

Options: Multiple, Polyline, Radius, Trim mode, Undo or

Specify first entity» P1

Options: Shift + select to apply corner or

Specify second entity» P2

: FILLET

Mode = TRIM, Radius = 0.5

Options: Multiple, Polyline, Radius, Trim mode, Undo or

Specify first entity» P3

Options: Shift + select to apply corner or

Specify second entity» P4
```

The `Multiple` option doesn't end the command after selecting two entities; thus allowing for multiple applications. The `Polyline` option fillets all the vertices of a polyline. The `Radius` option allows the user to modify the default radius. The `Trim` option allows the user to extend or trim entities in order to match endpoints of the created arc, or doesn't modify original entities and just adds the arc. The `Undo` option, when in `Multiple` mode, undoes the last fillet.

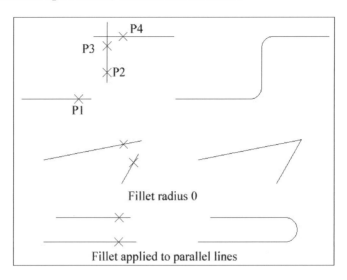

Fillet radius 0

Fillet applied to parallel lines

 If you press the *Shift* key while selecting the second entity, a 0 radius is applied and the corner becomes acute.

The CHAMFER command (alias cha, ▢ on the **Modify** toolbar, or **Modify** main menu) allows the user to cut or chamfer corners. By default, it prompts only for the selection of two entities and applies the default distances:

```
: CHAMFER
(Trim mode) Active chamfer Dist1 = 1, Dist2 = 0.5
Options: Angle, Distance, mEthod, Multiple, Polyline, Trim mode, Undo or
Specify first line» P1
Options: Shift + select to apply corner or
Specify second line» P2
```

Most options are similar to FILLET. The Multiple option doesn't end the command after selecting two entities, thus allowing for multiple applications The Polyline option cuts all vertices of a polyline. The Trim option allows the user to extend or trim entities in order to match endpoints of the created segment, or doesn't modify original entities and just adds a segment. The Undo option, when in Multiple mode, undoes the last cut. The Distance option allows the user to modify both default distances. The Angle option allows defining the cut with a distance and an angle from the first entity, instead of two distances. The Method option allows the user to select the default method for cuts whether it is if two distances or distance and angle.

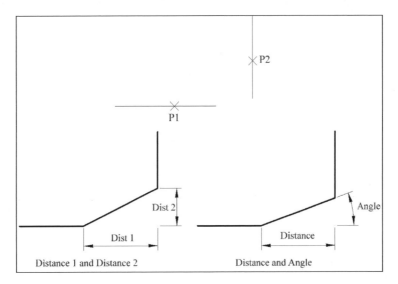

If you press the *Shift* key while selecting the second entity, distances with 0 values are applied and the corner becomes acute.

Exercise 3.1

We are starting to create a mechanical project with two views. This project will continue in upcoming chapters.

1. With the NEW command, start a new drawing from the standardiso.dwt template.

2. With the LINE command and the Ortho or Polar functions, draw the two views displayed in the following screenshot:

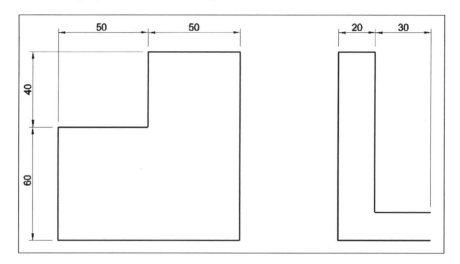

3. Apply the OFFSET command three times: the first with a distance of 15 applied to the bottom line, the second 5 applied twice to the line created with the first offset, and the third 30 applied to exterior vertical lines.

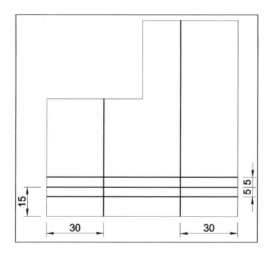

4. Apply the TRIM command to clean these lines. Press *Enter* to select everything as cutting edges and then cut all lines as displayed in following screenshot.

5. Apply the CIRCLE command-to define the center from the upper left corner the **From** entity snap may be used (for instance, selected from the *Ctrl* + mouse right button menu), picking the corner as a base point and @40,-15 as offset; the circle radius is 7.5.

6. To create the top inner line, apply the LINE command again. The first point is obtained with the **From** entity snap, picking the top-left corner as base and typing @15,-15 as offset. Then direct the cursor to the right and type 20.

7. Apply the OFFSET command to the last line, a distance of 10 down.

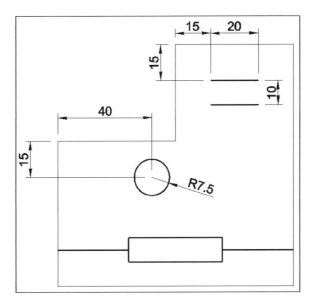

8. We want to cap the inner parallel lines with 180⁰ arcs. The easiest process is to apply the FILLET command in each side. There is no need to change the default radius.

9. Repeat the FILLET command to round the two top corners, but now applying the Radius option to modify default radius to 15.

10. Apply the CHAMFER command to cut the top-left corner. With the Distance option specify default distances of 20 and 25, and select the horizontal line first.

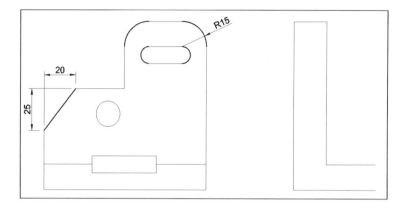

11. Save the drawing with the name PROJECT1-CH3.DWG.

 We could draw several additional lines and later move them to new layers that will be created in the next chapter. But the most efficient process is to create them already within the correct layers. Also the left view (on the right) could have more information, but we wait until almost everything is done and only then apply the MIRROR command.

Summary

In this chapter we presented the most important commands to start creating projects, including different types of entities creation and modification.

Adding to lines in the last chapter, the creation of circles, arcs, rectangles and regular polygons were included. When selecting entities, six more methods were introduced, adding flexibility to the selection process. Then we saw how to move and copy entities, how to apply reflections in order to obtain symmetries, how to rotate and apply a scale factor to entities, how to create parallel entities, how to move vertices, thus allowing for stretching entities, how to trim and extend entities, and how to apply rounds and diagonal cuts to corners.

In the next chapter, we will present entities properties, including layers, colors, types of lines, and lineweights.

4

Structuring Projects and Following Standards

Until now, everything has been created on layer 0, with white color and continuous linestyle. One of the great advantages of CAD is the ease in structuring all the project information and managing layers is a key factor in this. By dividing entities into layers, entities constitute mechanical pieces, nuts, bolts, sections, and so on.

In this chapter we will cover the following topics:

- What is a layer
- Creating and managing layers
- Assigning colors
- Loading and assigning linestyles
- Assigning lineweights
- Transferring entities from one layer to another

Creating and managing layers

A layer can be understood as a named transparent sheet. Overlapping all layers is the project. Each layer has several properties like visibility, locked, color, linestyle, lineweight, or printed.

There is a special layer named 0 (zero), existing in all drawings and that can not be deleted or renamed.

When creating entities, layer 0 should be avoided. However, this layer is useful when creating entities that will belong to components (blocks).

The layer list

To activate a layer and control the main layer properties the quickest process is applying the layer list, contained in the **Layers** toolbar.

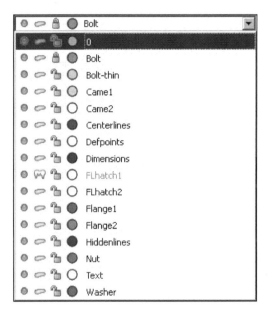

This list can easily be used for the following operations:

- **Activate a layer**: To activate a layer (make it current for drawing next entities) its name must be picked with the cursor.

- **Modify layer entities**: If we want to move entities from one layer to another, we select these entities, without command, and select the wanted layer in the list.

- **Turn layers on or off**: The first symbol, a circle, indicates if the layer is turned on (green) of off (grey).When a layer is turned off, all their entities are hidden, thus simplifying the project. However, these entities are processed by DraftSight, whenever a rebuild, zoom extents or select ALL is executed. It is also possible to activate a layer that is off, meaning that what is being created is immediately hidden.

- **Freeze or thaw layers**: The second symbol indicates if the layer is thawed (water) or frozen (ice). When a layer is frozen, all their entities are hidden and not processed by DraftSight. The active layer cannot be frozen.

> Despite entities being hidden by turning off or freezing layers, it is advisable to apply freezing. It is safer and quicker.

- **Lock or unlock layers**: The third symbol, a lock, indicates if the layer is locked (closed lock) or unlocked (opened lock). When a layer is locked, all their entities, while visible and selectable, can not be modified.

The LAYER command

The LAYER command (alias LA, on the **Layers** toolbar, or **Format** main menu) displays the **Layers Manager** dialog box for creating and managing layers.

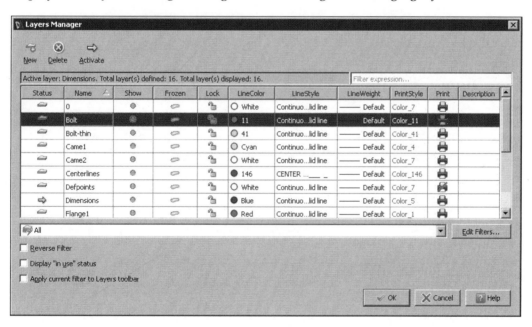

In this box, it is possible to select more than one layer. By pressing the *Shift* key, all layers between the selected one and the one being picked are selected. By pressing the *Ctrl* key, it is possible to select more layers, or remove them from selection.

The first button, **New**, creates a layer. A new line is added below the selected layer, with the **Layer1** (or next available number) name selected and ready to be modified. It inherits the properties of the selected layer. The second button, **Delete**, deletes selected layers, provided that they have no entities. The third button, **Activate**, activates the selected layer.

The layers list includes the following columns:

- **Status**: This column indicates which layer is current (arrow symbol) and, if the **Display "in use" status** option is checked, which layers have no entities (white symbol).
- **Name**: By picking twice quickly in a layer name, this can be modified.
- **Show**: This is the same as the on/off symbol in the layers list, meaning that layers can be turned on or off.
- **Frozen**: This is the same as the freeze/thaw symbol in the layers list, meaning that layers can be frozen or thawed.
- **Lock**: This is the same as the lock/unlock symbol in the layers list, meaning that layers can be locked or unlocked.

By applying the LAYER command, instead of the layers list, it is possible to modify these last three properties for several layers at once.

- **LineColor**: This column allows the user to modify the color for all entities in the selected layers. It includes a list with the basic colors and the **Specify Color** option that displays the **Line Color** dialog box, presented next.

Colors are very important in CAD. Despite allowing for a much better drawing interpretation, they also allow for configuring prints.

- **LineStyle**: This column allows the user to modify the linestyle, or the type of line, for all entities in the selected layers. It includes a list with the loaded linestyles and the **Other** option that displays the **Line Style** dialog box, presented next.
- **LineWeight**: This column allows the user to modify the lineweight (width of line, that may be used for printing), for all entities in the selected layers. It includes a list with all defined lineweights. This property is detailed next.

- **PrintStyle**: If the drawing is set to be printed with CTB files, this column only displays the layer color. If the drawing is set to be printed with STB files, it includes a list with print styles and the **Other** option that displays the **Print Styles** dialog box. Print styles, CTB and STB files are presented in *Chapter 10, Printing Efficiently*.

- **Print**: This column allows for the user to print/not print all entities in the selected layers. Some information (for instance, project notes, alternative decisions, and viewport boundaries) may be useful for displaying on screen but not in prints.

- **Description**: This allows the user to include a description for each layer. Often, the layer name is not enough for understanding its contents or purpose.

There are several recognized standards for layer names and layer properties. This is critical for exchanging drawings between different offices or areas. The most used are **BS-1192** (British Standards), **AIA Cad layer Guidelines** (United States) and **ISO 13567** (Europe). More information and links can be found at http://en.wikipedia.org/ wiki/CAD_standards. Even if an office does not want to conform to one of these standards, it is very important that internal standards should be defined as soon as possible.

Filtering layers

When dealing with a large number of layers, management can be hard. The textbox **Filter expression**, on the top right of the dialog box, allows the user to define temporary filters by name. Wildcards may be used: ? (question mark) for a single character or * (asterisk) for a sequence of characters, as displayed in following screenshot:

Status	Name △	Show	Frozen	Lock	LineColor	LineStyle	LineWeight	PrintStyle	Print
⊝	Flange1	◉	⟳	🔒	● Red	Continuo...lid line	—— Default	Color_1	🖶
⊝	Flange2	◉	⟳	🔒	● 42	Continuo...lid line	—— Default	Color_42	🖶

Active layer: Centerlines. Total layer(s) defined: 16. Total layer(s) displayed: 2.   *ang*

Additionally, it is possible to create permanent filters by clicking on the **Edit Filters** button. In the **Edit Layer Filters** dialog box, we may create a new group filter and add layers by selecting entities in the drawing area belonging to these layers. The list below columns, besides **All** and **All Used Layers**, also include all named filters. Checking the **Reverse Filter** option, all layers not included in the filter are displayed. Checking the **Apply current filter to Layers toolbar** option, the picked is also applied to the layer list.

Applying colors, linestyles, and lineweights

Associated to layers, or applied directly to entities, three very important properties to ease project understanding, visualization and printing quality, are color, linestyles and lineweights.

The **Properties** toolbar allows the user to apply explicit colors, linestyles and lineweights directly to entities, thus being these properties no longer controlled by layer.

 Colors, linestyles and lineweights should be controlled by layer and not explicitly. For instance, when applying an explicit color to entities, these may be confused with another layer.

Applying colors

When selecting a color, either in the **Layers Manager** dialog box or explicitly, the **Line Color** dialog box is displayed.

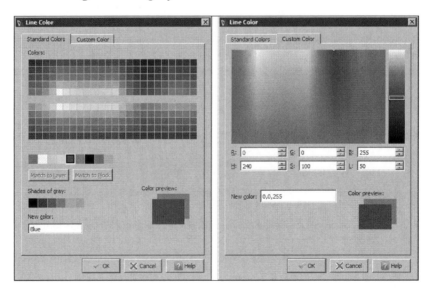

This box has two tabs:

- **Standard Colors**: This tab includes 255 standard colors. All colors are identified by a number, from 1 to 255. The first seven also are identified by name. For instance, 1 is red, 2 is yellow, 7 is white/black (depending on the background color).

- **Custom Color**: It is possible to assign a so-called **True Color**. Such color can be specified from RGB (red, green, blue) or HSL (hue, saturation, lightness) color models, assigning values from 0 to 255. These combinations result in 16.7 million colors available.

 To control printing with CTB plot style files, we can only assign printing settings to standard colors. Custom colors can not be controlled.

Applying linestyles

In technical drawings it is frequent to apply different linestyles or line types. For instance, center, or symmetry lines are represented by a sequence of long dash, small dash, and hidden lines by a sequence of equally sized dashes. In DraftSight, to apply a non-continuous linestyle, this must be available in the drawing; if not, it must be loaded.

When selecting a linestyle, either in the **Layers Manager** dialog box, in the **Properties** toolbar or by applying the LINESTYLE command (alias LT, **Format** main menu), the **Line Style** dialog box is displayed. This box includes all loaded linestyles. If the wanted linestyle is not available, by pressing the **Load** button, the **Load LineStyles** is displayed. The active linestyle can be modified by clicking on the **Browse** button.

Linestyles are defined in text files with the LIN extension. DraftSight includes two such files, MM.LIN and INCH.LIN, whose location can be checked with the OPTIONS command | **File Locations** area | **Drawing Support** | **LineStyle File**.

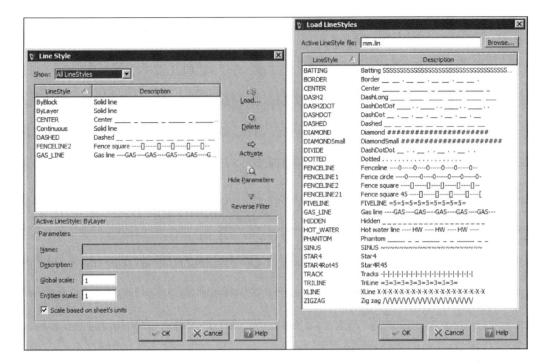

In the **Load LineStyles** box we select one or more linestyles and click on **OK**. The process can be repeated whenever new linestyles are needed.

Other buttons in the **Line Style** dialog box include:

- **Delete**: To delete a linestyle definition from the drawing
- **Activate**: To make a linestyle current, not by layer
- **Show Parameters**: To access additional parameters
- **Reverse Filter**: To display linestyles not belonging to the filter selected in the Show list

Additional parameters include modifying **Name** and **Description**, only in the current drawing, and also **Global scale** and **Entities scale**.

Linestyle elements, such as dashes, spaces, symbols or text, are defined with dimensions. Depending on the drawing dimensions, linestyles may be too small or too large. There are two possibilities for controlling linestyle scales:

- **Global scale**: With the **Global scale** textbox or by applying the LINESCALE command (alias LTS), a global scale factor for all linestyles is specified.
- **Entities scale**: It is possible to assign a different linestyle scale to selected entities with the PROPERTIES command, presented in the next chapter.

 Normally, modifying the **Global scale** factor is enough. All linestyles, even those not loaded yet, will adjust accordingly.

Applying lineweights

In mechanical drafting it is also important to control printed lineweights (sometimes described by width or thickness). DraftSight includes a complete set of predefined weights, as used in technical drawings, displayed when selecting a lineweight, either in the **Layers Manager** dialog box or in the **Properties** toolbar.

Besides printing, lineweights can also be displayed in the drawing area. To control this behavior, in the absence of a button on the Status bar, we must go to the OPTIONS command | **Drafting Styles** area | **Active Drafting Styles** | **Line Font** | **Line Weight**. A better method to reach these parameters is by applying the LINEWEIGHT command (alias LW, **Format** main menu).

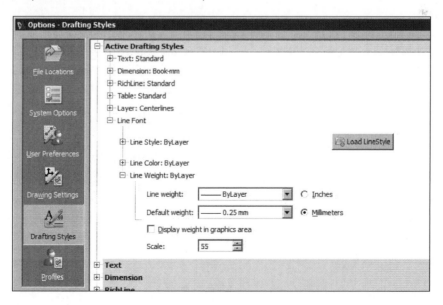

By checking **Display weight in graphics area**, lineweights equal or greater than 0.3 mm are displayed thicker.

Exercise 4.1

We are going to continue the mechanical project started in the last chapter, by creating layers, loading linestyles and moving entities to layers. This drawing needs eight layers and two linestyles. As our project is evolving, it is important to notice that the drawing on the left will be a section defined on the right drawing and we are using First Angle Projection.

1. Open the drawing PROJECT1-CH3.DWG.

2. Apply the LINESTYLE command (alias LT), click on the **Load** button, select the **HIDDEN** and **CENTER** linestyles (*Ctrl* + click to select more than one linestyle) to load them into the drawing and click on **OK**. Click on **OK** again to exit the **Line Style** dialog box.

3. Now, apply the LAYER command (alias LA) to create the eight layers and assign linestyles and lineweights as displayed in the next image. Colors can be at choice, depending on the background color. Activate the **HIDDEN LINES** layer and press **OK** to exit.

Active layer: HIDDEN LINES. Total layer(s) defined: 9. Total layer(s) displayed: 9.

Status	Name	Show	Frozen	Lock	LineColor	LineStyle	LineWeight	Pr
	0	◉	�	🔒	○ White	Continuous Solid line	—— Default	No
	CENTER LINES	◉		🔒	○ Green	CENTER ...___ _	—— 0.13 mm	No
	CUTTING PLANE	◉		🔒	● 190	CENTER ...___ _	—— 0.35 mm	No
	CUTTING TEXT	◉		🔒	● Magenta	Continuous Solid line	—— 0.25 mm	No
	DIMENSIONS	◉		🔒	● 24	Continuous Solid line	—— 0.13 mm	No
	HATCH1	◉		🔒	● Blue	Continuous Solid line	—— 0.13 mm	No
	HATCH2	◉		🔒	○ Cyan	Continuous Solid line	—— 0.13 mm	No
⇨	HIDDEN LINES	◉		🔒	● Red	HIDDEN ..._ _ _ _	—— 0.25 mm	No
	OUTLINE	◉		🔒	○ White	Continuous Solid line	—— 0.35 mm	No

4. Apply the LINEWEIGHT command (alias LT) and check the **Display weight in graphics area** option.

5. Select all entities, without command, and in the layer list pick the **OUTLINE** layer. All entities should be a little bit thicker.

6. With the **HIDDEN LINES** layer current, apply the LINE command and draw all four hidden lines on the right part of the drawing. **Polar, ESnap** and **ETrack** auxiliary tools should be on, so projecting holes is easier.

7. Hidden lines should be smaller. Apply the LINESCALE command (alias LTS) and specify a global scale of 0.3.

8. There are two edges missing, corresponding to the projection of the chamfer application. Activate the OUTLINE layer, and apply the LINE command to draw the two lines.

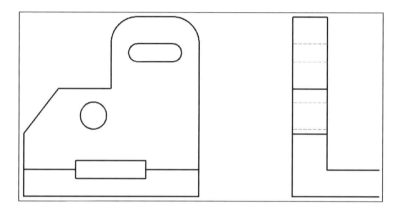

9. Now, the center lines. Activate the **CENTER LINES** layer. Draw the five lines in the left view and the two lines in the right view, applying the auxiliary tools.

10. Activate the **CUTTING PLANE** layer and draw the symmetry line.

11. Apply the MIRROR command (alias MI) and complete the right view.

12. To draw the inner part (pin), activate layer **OUTLINE**. Apply the LINE command, pause over the point marked as A, direct the cursor left and type 10, direct up and type 5, direct right and type 20, direct down and type 5.

13. Apply the MIRROR command (alias MI) to these three lines.

14. Select the three lines coming from the MIRROR and, in the layer list, specify the **HIDDEN LINES** layer.

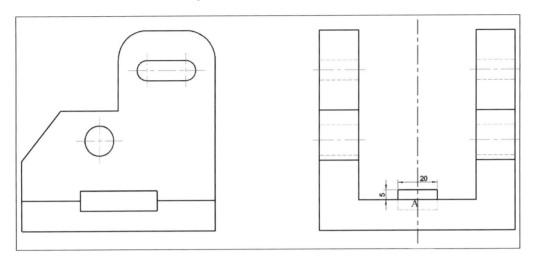

15. Save the drawing with the name PROJECT1-CH4.DWG.

Summary

This chapter includes the main entity properties, essential to a correct structure of any drawing, being layer the most important. Any entity must belong to a layer and overlapping all layers is the drawing. Each layer has several properties like visibility, locked, color, linestyle, lineweight, or printed.

The layer list controls the most used properties, including activating and modifying entities layer. The LAYER command includes some common properties and some others.

Associated to layers, or applied directly to entities, three very important properties to ease project understanding, visualization and printing quality, are color, linestyles and lineweights. These properties should be controlled by layer. Concerning linestyles, it is possible to control global scale (LINESCALE command), applied to all linestyles, or entity scale.

In the next chapter we will see how to calculate distances, areas and obtain other information from the project, as well has modifying one or more entities properties with a single operation.

5

Inquiring Projects and Modifying Properties

A common need in CAD is to obtain information in the drawing, such as verifying a distance or calculating an area. This chapter includes two commands to modify entities properties and a command to select entities based on properties values.

In this chapter we will cover the following topics:

- Measuring distances and angles between two points
- Calculating areas
- Obtain absolute coordinates for points
- Verifying and modifying generic properties
- Verifying and modifying specific entities properties
- Selecting entities according to properties values
- Copying properties between entities

Measuring distances and angles

The GETDISTANCE command (alias DI, on the **Inquiry** toolbar, or **Tools | Inquiry** main menu) calculates the distance, increments along the three directions, and absolute angles, in XY plane and from XY plane. The command just prompts for picking two points:

```
: GETDISTANCE
Specify start point»P1
Specify end point»P2
Distance = 19.447, Angle in XY Plane = 15, Angle from XY Plane = 0
Delta X = 18.793, Delta Y = 5, Delta Z = 0
```

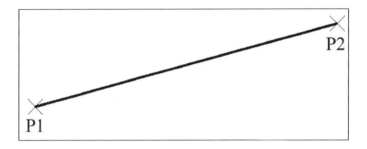

[This is one of the most useful commands, either to verify if the drawing is correct or obtaining information leading to costs estimation.]

Measuring areas

The GETAREA command (alias AA, on the **Inquiry** toolbar, or **Tools | Inquiry** main menu) calculates areas and perimeters. By default, the command just prompts for picking points, *Enter* to end the command and the area and perimeter are displayed:

```
: GETAREA
Options: Add, specify Entity, Subtract or
Specify first point» P1
Options: Enter for total or
Specify next point» P2
Options: Enter for total or
Specify next point» P3
Options: Enter for total or
```

```
Specify next point» P4
Options: Enter for total or
Specify next point» P5
Options: Enter for total or
Specify next point» P6
Options: Enter for total or
Specify next point» Enter
Area = 214, Perimeter = 70
```

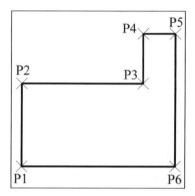

The Add and Subtract options allow calculating areas by adding and subtracting areas without leaving the command. The specify Entity option allows user to select entities whose types include area as property, like circles, arcs, polylines or ellipses.

> When trying to make calculations, the Add option must be the first option to apply.

The following diagram simulates a cut piece, where the outer line is a polyline. The method to calculate its area is as follows:

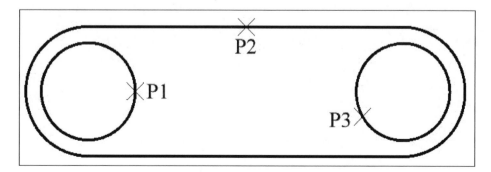

```
: GETAREA
Options: Add, specify Entity, Subtract or
Specify first point» A (Add option)
Options: specify Entity, Subtract or
Specify first point» E (specify Entity option)
Add mode...
Specify entity» P1
Area = 210.265, Perimeter = 65.133
Total area = 210.265
Add mode...
Specify entity» Enter
Options: specify Entity, Subtract or
Specify first point» S (Subtract option)
Options: Add, specify Entity or
Specify first point» E (specify Entity option)
Subtract mode...
Specify entity» P2
Area = 28.274, Perimeter = 18.85
Total area = 181.991
Subtract mode...
Specify entity» P3
Area = 28.274, Perimeter = 18.85
Total area = 153.717
Subtract mode...
Specify entity» Enter
Options: Add, specify Entity or
Specify first point» Enter
```

Obtaining coordinates

The GETXY command (alias ID, ▣ on the **Inquiry** toolbar, or navigating to
Tools | Inquiry | GetCoordinate) allows user to obtain the absolute coordinates
of points. By default, the command just prompts for picking a point:

```
: GETXY
Specify location» Point
X = 655.557 Y = -249.509 Z = 0
```

Verifying and modifying properties

The GETPROPERTIES command (alias LI, ▣ on the **Inquiry** toolbar, or navigating to **Tools | Inquiry**) lists entities' properties. It includes entities' type; generic properties such as layer, line color, linestyle and scale, and lineweight; and specific properties, depending on entities' type, as center or endpoints coordinates, radius, text height, and so on. The command just prompts for entities selection, and their properties are displayed:

```
: GETPROPERTIES
Specify entities» P1
1 found, 1 total
Specify entities» P2
1 found, 2 total
Specify entities» P3
1 found, 3 total
Specify entities» Enter
LWPOLYLINE
Handle: 3C7
Mode: Model
Layer: thick
LineColor: ByLayer
LineStyle: ByLayer
LineWeight: ByLayer
PolyLine Flags: Closed
Area: 130.265
Perimeter: 45.133
Location 0: X=668.591 Y=-286.217 Z=0
Location 1: X=678.591 Y=-286.217 Z=0
Location 2: X=678.591 Y=-278.217 Z=0
Location 3: X=668.591 Y=-278.217 Z=0
CIRCLE
Handle: 3C8
Mode: Model
Layer: thick
LineColor: ByLayer
LineStyle: ByLayer
LineWeight: ByLayer
```

```
Center point: X=658.45 Y=-282.217 Z=0

Radius: 4

Circumference: 25.133

Area: 50.265

LINE

Handle: 3C6

Mode: Model

Layer: thick

LineColor: ByLayer

LineStyle: ByLayer

LineWeight: ByLayer

from point: X=648.855 Y=-286.217 Z=0

to point: X=653.355 Y=-278.423 Z=0

Length: 9

Angle in XY Plane: 60

Delta X: 4.5

Delta Y: 7.794

Delta Z: 0
```

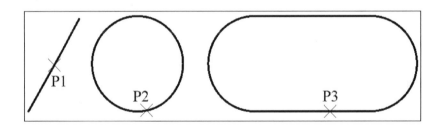

The result of this command is displayed on a textbox, representing the command line history. The *F2* key allows the user to turn this textbox on or off.

Unlike other DWG software, in DraftSight the order of displaying the information is opposite to the selection order.

The PROPERTIES command (shortcut *Ctrl + 1*, ▦ on the **Standard** toolbar, or the **Modify** main menu) allows user to verify and modify entities properties. It displays a palette providing access to all generic and geometric properties of selected entities. This palette can be always visible, docked to a side, or floating.

The list on the top of the palette indicates how many objects are selected.

If no entity is selected, the palette includes all generic properties, such as layer, line color, linestyle and scale, and lineweight for next entities to be drawn. Also included are default print style, drawing area information and coordinate system icon properties.

When selecting a single entity, all generic and geometric properties for that entity are included. Fields with gray background are only informative.

This palette also allows user to select multiple entities. If all entities are of the same type, geometric properties are displayed and may be modified. Selecting entities of different types, only generic properties are displayed. However, by filtering the list on the top of the palette, it is possible to select a specific type.

The palette includes three buttons: the first commutes between adding to selection or replacing selection; the second prompts for selection on the command line; the third applies the SMARTSELECT command, as shown in the following screenshot:

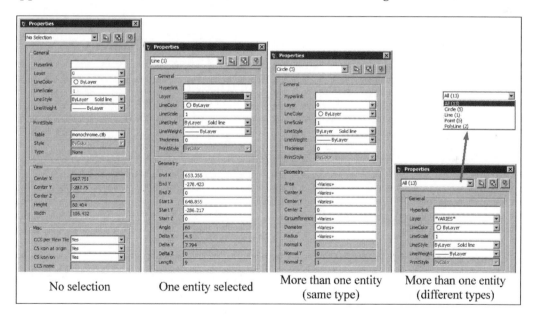

| No selection | One entity selected | More than one entity (same type) | More than one entity (different types) |

 Ctrl + 1 or the toolbar icon are easy processes for displaying or hiding this palette.

The SMARTSELECT command (on the **Properties** palette) allows user to create a selection by filtering entities properties.

The command displays the dialog box displayed in next image. In **Apply to**, the filter can be applied to the entire drawing or to a selection, made before entering the command or by pressing the button on the right. In **Entity**, the filter can be applied to a specific type of entity or to multiple types. All properties are presented in the **Property** area; if **Multiple** is chosen, only generic properties are available. Depending on the selected property, **Operator** allows user to specify **= Equals, <> Not equal to, Specify All, > Greater than** or **< Less than**. **Value** allows 'user to specify a comparing value: typed if numeric or selected if named. In **Selection results**, **Add to Selection Set** creates the selection based on the filter, and **Remove from Selection Set** has the opposite effect, selecting all entities that are not included in the filter. The **Add to existing Selection Set** option allows the user to specify more than one filter, accumulating selection sets.

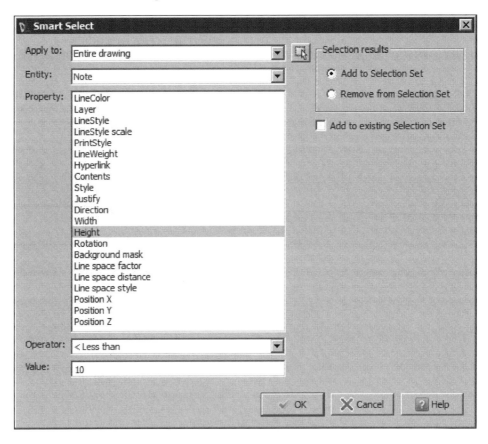

The PROPERTYPAINTER command (alias MA, 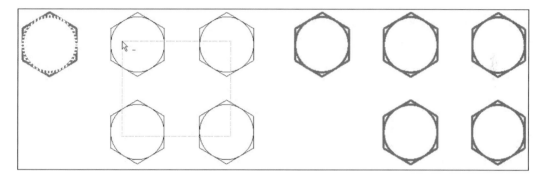 on the **Standard** toolbar, or **Modify** main menu) allows user to copy properties from a source entity to other entities. By default, the command prompts for selecting a source entity and, then, destination entities:

```
: PROPERTYPAINTER
Specify source entity» Selection
Options: Settings or
Specify destination entities» Selection
8 found, 8 total
Options: Settings or
Specify destination entities» Enter
```

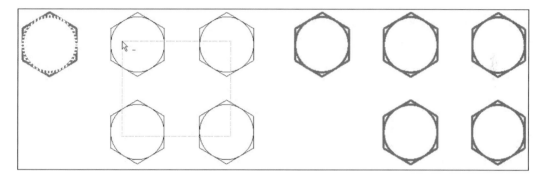

The Settings option displays a dialog box to select which properties are going to be copied. Besides all generic properties, being **Layer** the most used, the command can also copy specific properties of text, dimensions, hatches, tables and viewports.

Exercise 5.1

We are going to apply several commands presented in this chapter to a metal plate with several bolts:

1. Open the drawing EXERCISE-CH5-START.DWG.

2. Measure the diameter of the outer circle, by applying the GETDISTANCE command (alias DI) and selecting two opposite quadrants. Result should be 200. Another possibility could be applying the GETPROPERTIES command (alias LI).

3. Select, without command, the inner circle. Apply the PROPERTIES palette (*Ctrl + 1*) and modify the radius to 30.

4. We want the area of the metal plate, that is, between the outer and inner circles. Apply the GETAREA command (alias AA), set the **Add** option, **Entity** option and select the outer circle. The command displays: Area = 31415.927, Perimeter = 628.319, Total area = 31415.927. Press *Enter* to exit entity selection, set the **Subtract** option, **Entity** option and select the inner circle. The command displays: Area = 2827.433, Perimeter = 188.496, Total area = 28588.493. Press *Enter* twice to exit the command.

5. Select, without command, one hexagon and its inner circle. Apply the PROPERTIES palette (*Ctrl + 1*) and modify the layer of these two entities to **Bolt**.

6. Finally, applying the PROPERTYPAINTER command (alias MA), select the hexagon or the circle moved to the **Bolt** layer and apply its properties to all hexagons and respective circles.

7. Save the drawing with the name EXERCISE-CH5.DWG.

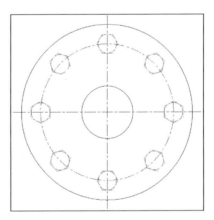

Summary

This chapter includes several commands for obtaining information in the drawing, two commands to modify entities properties and a command to select entities based on properties values.

The GETDISTANCE command calculates distance, increments and angles between two points. The GETAREA command calculates areas and perimeters. The GETXY command allows user to obtain the absolute coordinates of points. The GETPROPERTIES command lists entities properties. The PROPERTIES command displays a palette that can be permanently visible, providing access to all generic and geometric properties of selected entities. From this palette we can apply the SMARTSELECT command for creating a selection by filtering entities properties. The PROPERTYPAINTER command allows user to copy properties from a source entity to other entities.

In the next chapter we will see some more commands to create complex projects, as polylines, ellipses, revision clouds, text, tables, and multiple equally-spaced copies.

6

Creating Complex Projects

This chapter includes several commands and concepts that allow us to create more complex projects such as text, polylines, tables, ellipses, and other entities. Many other additional editing commands are also presented.

The following topics are covered in this chapter:

- Creating and managing text styles
- Creating single-line and multiline text
- Creating polylines without previous lines and arcs
- Converting continuous lines and arcs into polylines
- Creating polylines from boundaries of closed areas
- Creating multiple equally-spaced copies
- Creating ellipses
- Creating rings and revision clouds
- Creating tables and table styles
- Welding entities
- Modifying the length of entities
- Breaking entities and exploding complex entities

Creating text styles and applying text

In drafting, it is often necessary to add notes and text. DraftSight includes a command to create and manage text styles, and two commands that add notes to the drawings.

The TEXTSTYLE command (alias ST, 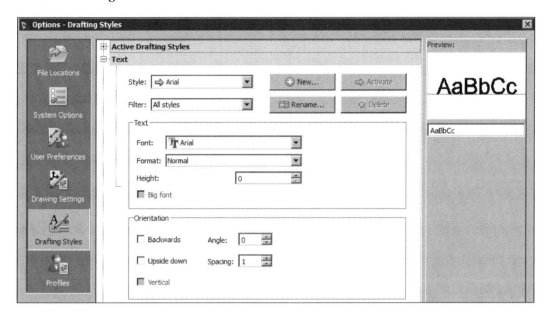 on the **Text** or **Styles** toolbars or on the **Format** main menu) creates and manages text styles. This command directly opens the **Text** area of the **Drafting Styles** section within the **Options** dialog box, as shown in the following screenshot:

The **Style:** list includes all the text styles defined in the drawing or that are being used by selecting **Styles in Drawing** in the **Filter:** list. The following are the four buttons to manage text styles:

- **New...**: This button is used to create a new text style, which you can name yourself
- **Activate**: This button is used to specify the listed text style as default to the new texts
- **Rename...**: This button is used to rename a text style
- **Delete**: This button is used to eliminate the text style if it is not used

To the listed style, it is possible to select a **Font**, its **Format** (between **Normal, Bold, Italic,** and **Bold Italic**), a fixed **Height,** and whether it uses a **Big font** (extended oriental characters). The **Orientation** section controls additional options for the default text style. The frequently used options are **Angle:** – to tilt the letters related to vertical and **Spacing:** – to shorten or lengthen the letters.

When not creating the text styles, DraftSight by default applies a text style named **Standard**, which cannot be deleted or renamed. A text height other than 0 will use this value for the next text commands.

The NOTE command (alias N or T, 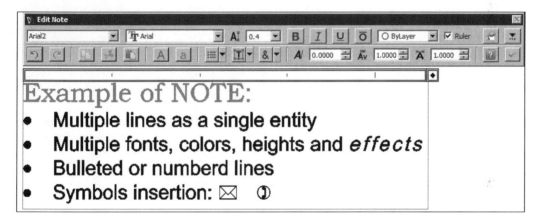 on the **Draw** or **Text** toolbars or on the **Draw | Text** main menu) creates multiline formatted text. This command prompts for two points defining the text width, and displays a rectangle with the text cursor inside along with a specific toolbar with two rows.

```
: NOTE
Active TextStyle: "Arial2" Text height: 0.51
Specify first corner» Point
Options: Angle, Height, Justify, Line spacing, textSTyle, Width or
Specify opposite corner» Point
```

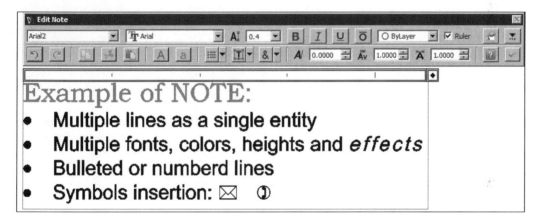

Most of the options are similar to any text processor. After typing and formatting the text, a single click outside or on the green checkmark icon ends the command.

The top row includes the following options: Style list, which is applied to all the text; Font list, which is applied to the selected text; Height, which is applied to the selected text. Then there are the Bold, Italic, Underline, and Overline buttons; Color list; Ruler display; Insert Field and Other Options buttons.

The bottom row of the preceding screenshot includes the following options: Undo and Redo buttons; Copy, Cut, and Paste buttons; Uppercase and Lowercase buttons; Bullets and Lists button; Alignment and Insert Symbol buttons; Angle, Tracking factor, and Spacing values; Help and Create Note buttons.

> The advised process to apply this command is to type everything, or paste it from an outside source, and then select and format the text as desired. To edit a note entity, we have to just click on it twice. The same editor that is used to create it is displayed.

The SIMPLENOTE command (alias DT, on the **Draw** or **Text** toolbars or **Draw | Text** main menu) creates single lines text. By default, it prompts for the start point of the text (left justification), height, angle, and text. After typing first line of the text, if you press *Enter*, the cursor moves to the next line, allowing typing of text below the previous line. Pressing *Enter* twice exits the command. It is also possible to click anywhere on the screen and the cursor moves to that position, allowing you to type text in that position. While typing, the right button menu includes the **Editor Settings | Show Toolbar** to display a toolbar with the text properties.

```
: SIMPLENOTE
Active TextStyle: "Arial2" Text height: 3
Options: Settings or
Specify start position»P1
Default: 3
Specify height» Value
Default: 0
Specify text angle»Value
Specify text» Example of Enter
Specify text» simple notes
P2
Specify text» with DraftSight
Specify text» Enter
Specify text» Enter
```

Example of
simple notes   with DraftSight

The **Settings** option displays a dialog box to set style, height, angle, and justification. When starting this command, it is possible to select a different justification, as displayed in the next screenshot. The **Align** and **Fit** justifications specify a complete justification, prompting for left and right points; **Align** adjusts height maintaining the width factor while **Fit** adjusts the width factor maintaining height.

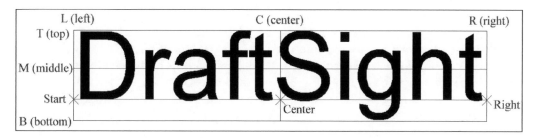

 The DT alias is actually for the SIMPLENOTE command. In this command, the **Settings** dialog box is replaced by the **Style** and **Justification** options, with the equivalent results.

Creating polylines

A polyline is a continuous sequence of lines and arcs that constitutes a single entity. There are three commands to create polylines. The RECTANGLE and POLYGON commands that we saw in *Chapter 3, Starting to Create Projects*, and the RING command that we will see later in this chapter also creates polylines but with specific shapes.

 Polylines are very useful when we need to calculate areas and lengths on the project, and for distributing points or components evenly spaced or at measured intervals, along its length.

The POLYLINE command (alias PL, on the **Draw** toolbar, or on the **Draw** main menu) creates polylines without existing entities. By default, this command prompts for the user to specify the points and draws straight lines between them; pressing *Enter* or *Esc* ends the command.

```
: POLYLINE
Options: Enter to continue from last point or
Specify start point» P1
Options: Arc, Halfwidth, Length, Undo, Width, Enter to exit or
Specify next vertex» P2
Options: Arc, Close, Halfwidth, Length, Undo, Width, Enter to exit or
Specify next vertex» P3
Options: Arc, Close, Halfwidth, Length, Undo, Width, Enter to exit or
Specify next vertex» P4
Options: Arc, Close, Halfwidth, Length, Undo, Width, Enter to exit or
Specify next vertex» Enter
```

The Halfwidth and Width options allow the user to specify half or full width for the next elements of the polyline, prompting for starting and ending values. The Length option allows the user to define a length for the next element maintaining tangency at the last point. The Undo option undoes the last element. The Arc option allows the user to include arcs in the polyline; a new set of options, similar to the ARC command, replaces the line mode options, the Line option coming back to this mode.

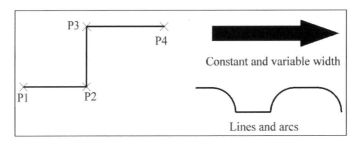

Constant and variable width

Lines and arcs

The EDITPOLYLINE command (alias PE, 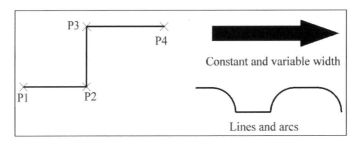 on the **Modify** toolbar or **Modify | Entity** main menu) edits polylines and creates polylines from the existing consecutive lines and arcs. This command prompts for the selection of a polyline. If you select a line or arc, the command asks if this is to be converted to a polyline. Then, it displays a set of options; pressing *Enter* exits the command.

```
: EDITPOLYLINE
Options: Multiple or
Specify polyline» Selection
Entity selected is not a PolyLine.
Default: Yes
Confirm: Do you want to turn it into one?
Specify Yes or No» Yes
Options: Close, Decurve, Edit vertex, Fit, Join, Linegen, Spline, Undo,
Width or eXit
Specify option» Enter
```

The Multiple option allows the user to apply editing options to several polylines at once. The Close or Open options close or open the polyline, depending on the existing state. The Join option allows the user to add other continuous lines or arcs (with coincident endpoints). The Decurve option removes all the arc elements of the polyline. The Edit vertex option includes several options for editing vertices or elements. The Fit and Spline options turn the polyline into another that are only composed by arcs; Fit polyline passes through all previous vertices; Spline polyline is a curved approximation. The Linegen option controls the behavior of non-continuous linestyles around vertices. The Width option allows the user to apply a uniform width to the polyline.

The most important application for this command is to turn a set of consecutive lines and arcs into a polyline by selecting one of the elements, confirming the polyline creation and then applying the `Join` option to select the remaining entities.

The AREABOUNDARY command (alias AB or BO, on the **Draw** main menu) creates polylines coincident with the boundaries of a closed area. This command displays a dialog box. The **Find nested boundaries** option, besides the external polyline also creates polylines coincident with the internal closed boundaries. In the **Performance** area it is possible to select which entities are verified for searching closed boundaries. The **Type** list allows the user to create polylines or regions.

When you click on the **OK** button, the command prompts for internal points and create the corresponding polylines in the active layer.

```
: AREABOUNDARY
"Insert Area Boundary" dialog box
Specify internal point» P1
Analyzing boundaries...
Specify internal point»
1 Boundaries were created
```

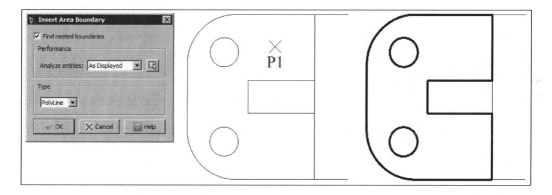

 When applying this command, its feedback may not be correct. In the preceding figure example, by picking the P1 point, three polylines are created (one exterior and two coincident with circles) and not one as informed by the command.

Creating multiple equally spaced copies

The PATTERN command (alias AR, ▦ on the **Modify** toolbar or on the **Modify** main menu) creates multiple equally spaced copies of the selected entities, with a rectangular or circular pattern. This command displays a dialog box where in the **Pattern type** area, a **Circular** or **Linear** (rectangular) pattern is specified. If it is not previously selected when entering the command, you can click on the **Select entities** button to access the drawing area to select entities.

When applying a **Linear** pattern, copies are made in rows and columns, whose numbers are indicated in **Number of elements on:**. Distances between equivalent points, in rows and columns, are indicated in **Spacing between elements on:**. With **Pattern angle** it is possible to define a **Linear** pattern rotated from horizontal. All the options are shown in the following screenshot:

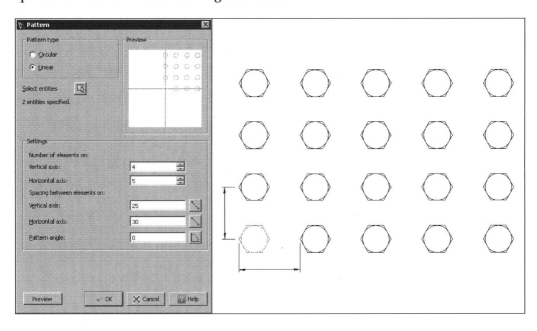

When applying a **Circular** pattern, copies are placed around a center point. There are three choices for distribution, on the **Base pattern on** list and they are: **Angle Between and Total Number of Elements, Fill Angle and Angle Between Elements,** and **Fill Angle and Total Number of Elements**. Depending on the distribution, two of the three text boxes are available: **Angle between, Fill angle,** and **Total number**. The **Axis point** area specifies the center of the circular pattern and the **Element base point** area specifies which point of the selected entities is used as reference. With **Orient elements about axis** checked, all copies are rotated, instead of the one just copied. All the options are shown in the following screenshot:

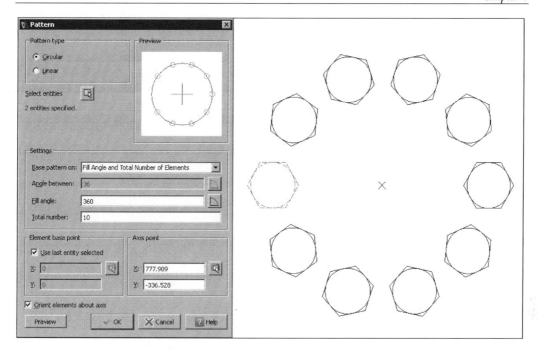

 After applying this command, there is no easy way to modify the number of circular elements or the number of rows and columns.

Creating other entities

There are some other entities that may be useful for technical drawings, namely ellipses, rings (donuts), and clouds (normally used for revisions).

The ELLIPSE command (alias EL, 🖉 on the **Draw** toolbar or on the **Draw** main menu) creates ellipses and elliptical arcs. By default, the command prompts for specifying two points that define the first axis and a point or a distance that defines the second half axis.

```
: ELLIPSE
Options: Elliptical arc, Center or
Specify axis start point» P1
Specify axis end point» P2
Options: Rotation or
Specify other axis end point» P3
```

The **Elliptical arc** option allows the user to create an elliptical arc by first defining the ellipse and then two points or angles for start and end of arc. The **Center** option allows the user to define an ellipse starting from its center and then two points for both half axes. The **Rotation** option creates an ellipse by projecting a circle rotated out of the plane, prompting for this rotation angle.

The RING command (alias DO or **Draw** main menu) creates rings and filled circles (if the inside diameter is 0). This command prompts for an inside diameter, outside diameter, and centers. Pressing *Enter* or *Esc* ends the command.

```
: RING
Default: 10
Specify inside diameter» Value
Default: 20
Specify outside diameter» Value
Default: Exit
Specify position» Enter
```

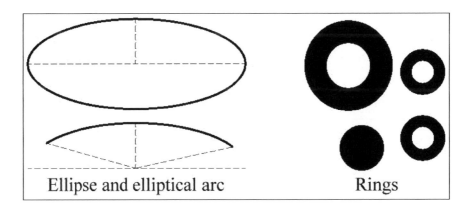

Ellipse and elliptical arc   Rings

The CLOUD command (no alias, ▧ ▧ ◌ ⟡ on the **Draw** toolbar or on the **Draw** main menu) creates revision clouds. These are unique entities and unlike other DWG software are not polylines. When applying the command, an option is requested.

```
: CLOUD
Radius = 1
Default: Rectangular
Options: Elliptical, Freehand, Rectangular, Radius or Settings
Specify option» Option
```

The **Elliptical** option creates a cloud with an elliptical shape. The **Freehand** option allows the user to create an irregular cloud by defining several segments. The **Rectangular** option creates a cloud with a rectangular shape. The **Radius** option defines the radius for arcs. The **Settings** option displays a dialog box for specifying all cloud parameters and properties; When you click on the **OK** button, a cloud with the selected shape is created as shown in the following figure:.

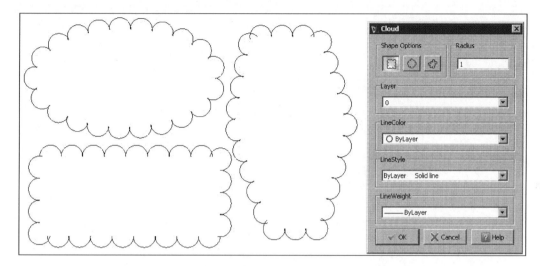

 These entities are useful for marking a part on the drawing that requires attention or needs revision. When exploding a cloud individual arcs are obtained.

Creating tables and table styles

Often there is the need to create tables in technical drawings, as bills of materials. A table is composed by cells and has a table style assigned.

The TABLE command (alias TB, on the **Draw** main menu) creates tables that are composed of rows and columns. This command starts by displaying a dialog box, which defines the number of rows and columns, also providing access to the TABLESTYLE command. After clicking on the **OK** button, the command prompts for the table position or for the opposite corners.

```
: TABLE
Specify first corner» Point
Specify opposite corner» Point
```

After specifying the corners or position, the first cell is selected and a textbox is displayed, allowing you to fill the cells. When the keyboard arrows or the *Tab* key is pressed, other cells can be accessed and fulfilled.

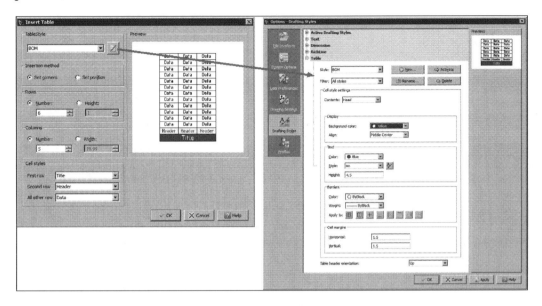

In the **TableStyle** area, an existing table style can be applied, or, by pressing the button, we access the TABLESTYLE command, to create or edit table styles. . The **Insertion method** area specifies whether the table is inserted into a position or defined by the two opposite corners with predefined parameters. The **Rows** area includes the number of data rows and height row (defined by text lines and not by dimension). The **Columns** area includes the number of columns and column width. The **Cell styles** area allows the user to specify which cell styles are applied to the **First row** (normally **Title**), **Second row** (normally **Header**), and **All other rows** (normally **Data**).

The EDITABLE command (no alias, on the **Modify | Entity** main menu) allows the user to modify the cell contents. This command prompts for the table selection and a cell selection, and displays a textbox with its content.

The TABLESTYLE command (alias TS, on the **Format** main menu) allows the user to create and modify table styles. This command displays the **Options** dialog box, the **Drafting Styles** section, and the **Table** area.

The **Style** list includes all the table styles defined in the drawing or that are being used, by selecting **Styles in Drawing** on the **Filter** list. The following are the four buttons to manage table styles:

- **New...**: This button is used to create a new table style, name being requested
- **Activate**: This button is used to specify the listed table style as default to the new tables
- **Rename...**: This button is used to rename a table style
- **Delete**: This button is used to eliminate the table style, if it is not used

In the **Cell style settings** area, we control, to each cell style selected in the Contents list (**Title, Header**, and **Data**), **Background color**, justification (**Align** list), **Text** properties, **Borders** properties, and **Cell margins**. The **Table header orientation** list controls whether the table is generated **Up** or **Down**.

With the DraftSight Version V1R3.1, it is not possible to create new cell styles and assign different formats to individual cells, like a background color.

Joining, lengthening, and breaking the entities

The following are few more editing commands that may ease several drawing tasks:

The WELD command (alias J, 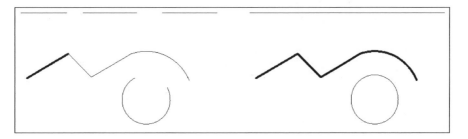 on the **Modify** toolbar or on the **Modify** main menu) allows the user to join entities. This command starts by prompting an entity Selection. Depending on its type, the command has several utilities.

```
: WELD
Specify base entity» Selection
```

When selecting a line, the command prompts for collinear lines and the result is a single line, removing breaks or overlaps. When selecting a polyline, the command prompts for other continuous entities and joins them to the polyline. When selecting an arc, besides joining other arcs with the same center and radius, it is possible to close the arc, transforming it into a circle. It is also possible to join elliptical arcs and splines.

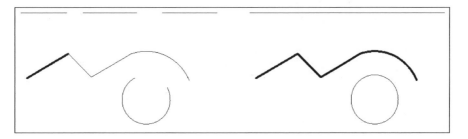

The EDITLENGTH command (alias LEN, navigate to **Modify | Change length** main menu) allows the user to verify and modify the length of the entities. By default, this command prompts for an entity selection and answers with its length (and included angle if it is an arc). The entity selection prompt is repeated until the *Esc* key is pressed.

```
: EDITLENGTH
Options: Dynamic, Increment, Percent, Total or
Specify entity for length» P1
Active length: 10
Options: Dynamic, Increment, Percent, Total or
Specify entity for length» P2
Active length: 5
Options: Dynamic, Increment, Percent, Total or
Specify entity for length» Esc
```

The **Dynamic** option allows you to modify the length dynamically by specifying the new endpoint with the cursor. The **Increment** option allows you to define an increment value; this value is added to the length and this value is subtracted from the length, if the value negative. The **Percent** option allows you to modify by percentage, adding length if more than 100 percent and removing if less. The **Total** option allows you to define the final length for entities to be selected. With the **Increment** and **Total** options, when selecting an arc, it is possible to modify its included angle.

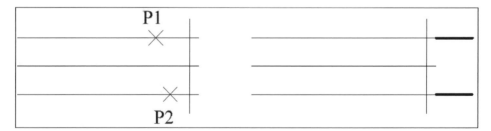

 The length modification is always performed from the endpoint nearest to the selection point.

The SPLIT command (alias BR, on the **Modify** toolbar, or on the **Modify** main menu) allows the user to erase part of the entities or to divide it into two connected entities. By default, this command prompts for an entity selection, using the selection point as first split point or a second split point:

```
: SPLIT
Specify entity» P1
Options: First point or
Specify second split point»P2
```

The **First point** option allows the user to specify a first split point just as an intersection entity snap, thus not applying the selection point. If the second point is coincident with the first point (for instance, @0, 0), an open entity is simply divided in that point.

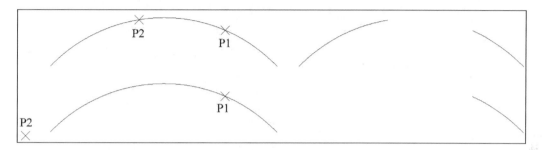

The EXPLODE command (alias X, 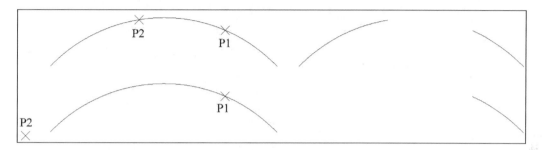 on the **Modify** toolbar, or on the **Modify** main menu) explodes composed entities such as polylines, notes, clouds, and components. The command just prompts for entities selection.

Exercise 6.1

Continuing the mechanical project from *Chapter 4, Structuring Projects and Following Standards,* we are going to measure the volume by applying text and creating a table. DraftSight, as other DWG softwares, is dimensionless, we assume the length unit is millimeters (mm)

1. Open the drawing PROJECT1-CH4.DWG.

2. To calculate the volume of both parts, the best way is to use a new layer to create a contour. Create a layer called **AUXILIARY**, with a different color and activate it.

3. With the POLYLINE command, create a closed polyline contouring the left view by applying the **Arc** option on the top fillets. The GETPROPERTIES command applied to this polyline indicates an area of 7653.429 mm2.

4. Apply the EDITPOLYLINE command to create a polyline from the two arcs and lines of the top hole. The GETPROPERTIES command of this polyline indicates an area of 278.54 mm2.

5. The circular hole area is `176.715`. So the total area is `7653.429-278.54-176.715`, that is, `7198.17`mm2. The width of each part is 20mm, meaning that the volume of both vertical parts is `7198.17 x 20 x 2`, which equals to `287926.8`mm2.

> Applying the `OSCALC` command, we can access the operating system calculator, which is handy for these calculations.
>
> DraftSight includes a scientific calculator, displayed with the `SMARTCALCULATOR` command (alias `QC`). This calculator is not precise, so it must be avoided. For instance, `7198.17 x 2` equals to `14396.34`, but the calculator only displays `14396.3`.

6. The central bottom part area is `100 x 60 x 15`, which equals to `90000`. The hollow for the pin is `40 x 20 x 5` which equals to `4000`. So the total volume of the main part is `287926.8 + 90000 - 4000` which equals to `373926.8` cubic units (mm3). The pin volume is twice the hollow which equals to `8000`.

7. Activate the **CUTTING TEXT** layer and freeze the **AUXILIARY** layer.

8. To add notes, a text style must be created. Apply the `TEXTSTYLE` command and create a style named `MODELTEXT`, with **FontISOCP**.

9. Apply the `SIMPLENOTE` command, with the text height as 5mm, and write `SECTION AA` above the left view.

10. This drawing should include a table with the volumes and material of both the parts. Apply the `TABLESTYLE` command and create a style named `MATERIALS`. All the cell styles should have a text style as `MODELTEXT` and **Cell margins** as 0.5; the **Data** and **Header** cell styles height as 3 and the **Title** cell style height as 3.5. Apply a different background color to the **Title** cell style.

11. Create a table with the TABLE command and fulfill it with the information displayed in the next table. The density of the C45E steel (1.1191) is 7.84 g/cm3.

Materials

Part nr..	Part name	Material	Volume (mm3)	Weight (Kg)
1	Support	Steel C45E	373926.80	2.932
2	Pin	Steel C45E	8000	0.063

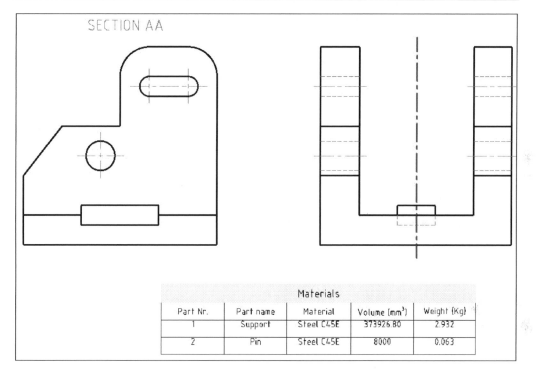

12. Save the drawing with the name EXERCISE-CH6.DWG.

Summary

This chapter includes several commands that allow specific functions that are important to create complex projects. The TEXTSTYLE command creates the text styles that are used by the NOTE and SIMPLENOTE commands; the first command creates multiline text while the second command creates a single-line text. To edit text you just have to double-click on it.

Polylines are a continuous sequence of lines and arcs that constitute a single entity. They can be created directly with the POLYLINE command by converting sequences of lines and arcs with the EDITPOLYLINE command or coincident with the boundary of a closed area with the AREABOUNDARY command.

The PATTERN command allows you to create multiple equally spaced copies of selected entities, with a rectangular or circular pattern. The ELLIPSE command allows you to create ellipses and elliptical arcs. The RING command allows you to create rings and filled circles. The CLOUD command allows you to create revision clouds. The TABLE command creates tables composed of rows and columns. The EDITABLE command allows you to insert and modify information inside table cells. Creating and managing table styles is done with the TABLESTYLE command.

Finally, a few more useful editing commands are included. The WELD command allows joining entities. The EDITLENGTH command allows you to verify and modify the length of entities. The SPLIT command allows you to erase a part of entities or dividing it into two connected entities. The EXPLODE command explodes the composed entities.

In the next chapter we will see how to create components (blocks), so we can manage libraries of common parts.

7

Creating and Applying Components

One of the great CAD advantages is the ability of reusing information. A component, or block, is a set of entities belonging to one or more layers, that are grouped together and that can be used as a single element. Examples are nuts, bolts, bearings, mechanical parts, and so on. When inserting components, besides position, scales (even different in X and Y directions) and rotation may be modified. Components can include variable text information, called attributes.

Topics covered in this chapter:

- Creating components
- Inserting components
- Exporting components
- Editing components
- Applying attributes
- Cleaning the drawing by removing unused definitions

Creating components (blocks)

The MAKEBLOCK command (alias B, on the **Draw** toolbar, or navigating to **Draw | Block**) creates a component from existing entities.

The command displays the **Block Definition** dialog box. In **Name** the component name is typed, or an existing component may be selected for overwriting it. In **Description**, additional information about this component may be typed in. In **Settings** the following options are available:

Apply uniform scale doesn't allow inserting this component with different scales; **Allow block to explode** indicates that this component may be exploded; **Attach Hyperlink** allows the user to assign a hyperlink to the component (for instance, site of the component manufacturer); **Units** list allows the user to associate a unit to the component. **Base point** allows the user to define the base point associated to selected entities; this point will be used to insert the component and can be specified in the drawing by the button or by typing its coordinates. **Block entities** allows the user to select entities that will make part of the component. Original entities can be individually preserved, converted to block or removed from the drawing. When you click on the **OK** button, the component is created.

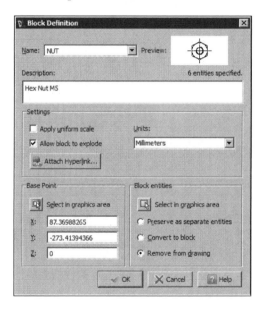

 The three most important component parameters are: name, base point (normally selected in the graphics area) and component entities. If the component is created without specifying a base point, the drawing origin is used as such, normally far away from the entities.

Inserting components

The INSERTBLOCK command (alias I, on the **Insert** toolbar, or navigating to **Insert | Block**) inserts a defined component in the drawing or a drawing file as a component.

This command displays the **Insert Block** dialog box. The **Name** list includes all components defined in the drawing that can be inserted. As an alternative, it is possible to insert a complete drawing as a component by pressing the **Browse** button. A file dialog box is displayed, in order to select a DWG or DXF file.

Position allows the user to specify where to locate the component by defining its X, Y and Z coordinates. **Scale** allows the user to define the scales in all directions. With **Apply uniform scale** checked, the component will be inserted with a uniform scale, without distortion. **Rotate** allows the user to apply a rotation to the component. With **Explode Block** checked, the component is exploded when inserted.

With **Specify later it is possible to define** position, scales and/or rotation after pressing **OK.** The following prompts are displayed:

```
: INSERTBLOCK
Options: Angle, reference Point, uniform Scale, X, Y, Z or
Specify destination» Point
Default: 1
Options: Corner, uniform Scale or
Specify X scale or specify opposite corner» Value
Default: 1
Specify Y-scale» Value
Default: 0
Specify angle» Value
```

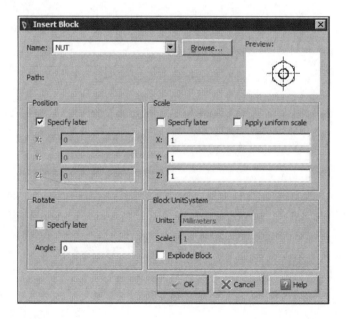

 When inserting a drawing, this becomes a component of the current drawing, thus enters the **Name** list.

Exporting components

The EXPORTDRAWING command (alias W, 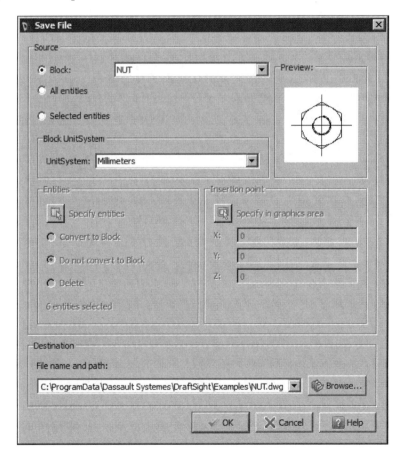 on the **Insert** toolbar, or navigating to **File | Export**) allows the user to export a component, all entities or just some entities, to a new DWG or DXF file.

The command displays the **Save File** dialog box. This box has two parts and a preview area. The **Source** area allows the user to specify what to export and settings. It is possible to export an existing **Block**, chosen on the list, **All entities** or just the **Selected entities**. The **UnitSystem** list allows the user to select the new drawing's unit.

With **Selected Entities checked**, in **Entities** we select the entities that will be exported and specify if these entities are converted to a block inside the current drawing, not converted or deleted. The **Insertion point** allows the user to define the base point.

In **Destination**, by clicking on the **Browse** button, the file format, version, and location can be specified.

This command can be used to separate a drawing into two or more drawings.

Editing components

The EDITCOMPONENT command (no alias, on the **Component** toolbar, or navigating to **Modify | Component**, or selecting a component and applying the mouse right button menu) allows the user to modify an existing component. The command starts by prompting for the selection of a component.

```
: EDITCOMPONENT
Specify component» Selection
```

Then, the **Edit Component** dialog box is displayed. The **Select** component area allows the user to specify an inner component to edit, if any. With **Zoom to Bounds** checked, the command applies a zoom to the component entities. **Show Selection** temporary hides the box to highlight the selected component. **Move Up** and **Restore** buttons allow modifying and restoring component's hierarchy. After clicking on **OK**, the drawing area gets a dark grey color, all entities fade except the entities from the component.

In this component environment, entities can be drawn, erased, or modified. The **Component** toolbar adds additional options: to transfer elements from the drawing to the component, to transfer elements from the component to the drawing, to modify the base point, to save the component and close this environment, or to close the environment.

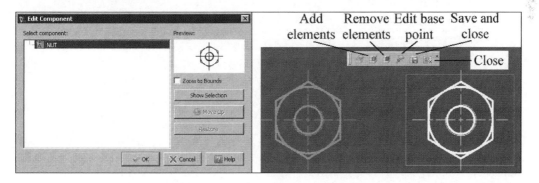

This command is also used to edit in-place external references. External references are presented in *Chapter 11, Advanced Tools*.

Applying attributes

Attributes are text information associated to components, such as reference, price, provider, or material. When inserting a component with attributes, their values are requested, thus allowing to associate useful information. Attributes are normally defined before creating the component.

The MAKEBLOCKATTRIBUTE command (alias ATT, no icon, **Draw | Block | Define Block Attributes**) creates attributes. It displays a dialog box, as shown in the following screenshot:

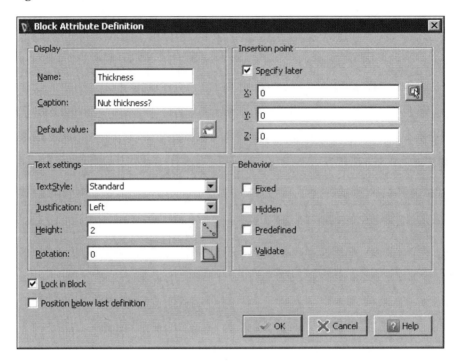

In **Display**, **Name** is the attribute name, identifying it, **Caption** is the question requesting an attribute value, when inserting a component, and **Default value** the value that can will be used as default.

The Insertion point defines the attribute position, normally near entities. The **Text settings** area includes the text properties for the attribute name and values, namely **TextStyle**, **Justification**, **Height** and **Rotation**.

The **Behavior** area includes some value properties: **Fixed** means that the attribute value is constant and uses the default value, it is not requested when inserting the component neither can be later modified; **Hidden** controls value visibility, if checked values are normally hidden; **Predefined** also applies default value but allows the user to modify the value later; **Validate** allows the user to confirm the value.

The **Lock in Block** option locks values, not allowing them to be moved by grip edit. **Position below last position**, available since the second attribute definition, places the attribute below the previous one.

To edit an isolated attribute, before creating the component, two easy ways are available. With an attribute selected, the PROPERTIES palette (*Ctrl + 1*) includes all parameters. Also the EDITBLOCKATTRIBUTEDEFINITION command (only typing its name) can be applied. This command displays the **Block Attribute Definition** dialog box, as presented before, but without the **Position below last position** option.

Components with attributes are created normally, with the MAKEBLOCK command (alias B). The insertion of a component with attributes is also with the INSERTBLOCK command (alias I).

After specifying an insertion point, eventually scales and rotation, all attribute values, except those that were set as fixed or predefined, are requested:

```
: INSERTBLOCK
Options: Angle, reference Point, uniform Scale, X, Y, Z or
Specify destination» Point
Enter BlockAttribute values
Default: M20
Hex nut size?» M18
Nut thickness?» 14.8
Default: 0.1
Nut price?» .15
```

By default, values are typed at the command line. But modifying the ENBLBATTDLGS variable (not documented, but ATTDIA can also be applied) from 0 to 1, values can be typed in the **Edit BlockAttribute Values** dialog box, as displayed in the following screenshot. This variable is recorded with DraftSight and normally there is no need to change it back:

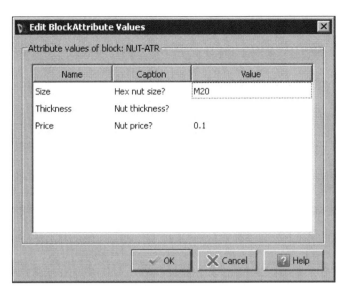

To modify attribute values, three options are available. When selecting a component with attributes, the PROPERTIES palette (*Ctrl + 1*) displays all values, allowing to modify those that are not fixed.

The EDITBLOCKATTRIBUTE command (can be accessed by navigating to **Modify | Entity | Block Attribute**, or double-clicking over a component with attributes) allows the user to modify attribute values, by displaying the **Edit BlockAttribute Values** dialog box.

Finally, the EDITIPBLOCKATTRIBUTE command (no alias, neither is available at the main menu) allows the user to edit attribute values in place. It just prompts for an attribute value selection.

Sometimes, we pretend to hide all attribute values, or see them all. The DISPLAYBLOCKATTRIBUTES command (**View | Display | Block Attribute**) controls attribute values visibility:

```
: DISPLAYBLOCKATTRIBUTES
Default: Normal
Options: Normal, OFf or ON
Specify option» Option
```

OFf option hides all attribute values, **On** displays all and **Normal** displays according to attribute definition.

Cleaning the drawing

When concluding a drawing, it is wise to clean the drawing by removing all unused definitions, such as layers, linestyles, text styles, table styles, components, and so on.

The CLEAN command (aliases CL or PU, **Format** main menu) allows the user to clean the drawing. It displays a dialog box including all definitions, divided by categories that can be removed. It is possible to make a full clean, or just remove selected definitions.

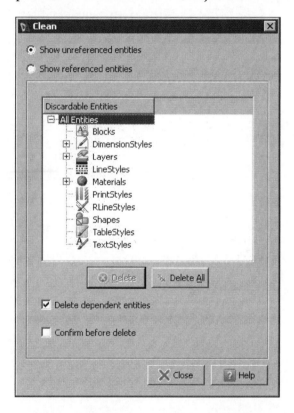

On the top, by selecting **Show referenced entities**, used definitions are displayed, instead of unused definitions. The **Delete** button removes selected definitions; the **Delete All** button removes all unused. With **Delete dependent entities** checked, inner definitions are also removed (for instance, components or layers defined only inside an unused component). With **Confirm before delete** checked, each removal requests confirmation on the command line.

Exercise 7.1

Continuing the mechanical project from last chapter, we are going to create a component with an attribute to number both parts.

1. Open the drawing PROJECT1-CH6.DWG.

2. Activate layer 0, if not active.

3. The component is simply a circle with an attribute inside. Create a circle somewhere in the drawing, with radius 6.

4. Create an attribute inside the circle (MAKEBLOCKATTRIBUTE command, alias ATT), **Name** NR, **Caption** Part number?, **Default value** 1, **Justification** Middle and **Height** 8. The **Insertion point** must be the center of the circle. In **Behavior**, all options must be unchecked.

5. When placing the attribute, maybe its height is too much. So select it and, with the PROPERTIES palette, modify the height to 6.

6. Select the circle and attribute. Apply the MAKEBLOCK command (alias B) to create a component with **Name** NUMBER, **Base Point** is the center of the circle (applying **Select in graphics area** button) and, for now, **Preserve as separate entities** selected. Click on **OK**.

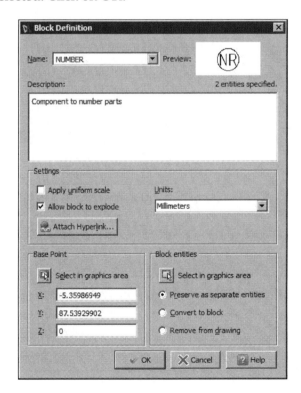

7. Just for testing, apply the INSERTBLOCK command (alias I) and insert the component somewhere. In **Position**, check **Specify later**. Confirm 1 as the attribute value.

8. Save the drawing with the name PROJECT1-CH7.DWG. This exercise continues in the next chapter.

Summary

This chapter includes commands related to the creation, insertion and other operations about components (blocks). A component, or block, is a set of entities belonging to one or more layers, that are grouped together and that can be used as a single element. To create components, there is the MAKEBLOCK command. The INSERTBLOCK command inserts a defined component in the drawing or a drawing file as a component. The EXPORTDRAWING command allows the user to export a component, all entities or just some entities, to a new DWG or DXF file. The EDITCOMPONENT command modifying an existing component, including defining a new base point.

Attributes are text information associated to components. When inserting a component with attributes, their values are requested, thus allowing the user to associate useful information. The MAKEBLOCKATTRIBUTE command creates attributes, their properties, as well as their values after component insertion, can be modified with the PROPERTIES palette. The DISPLAYBLOCKATTRIBUTES command controls attribute values visibility.

The CLEAN command allows the user to clean the drawing by removing all unused definitions, such as layers, text styles, components and so on.

In the next chapter we will see how to apply fills and patterns to closed areas.

8

Applying Fills and Patterns

In Mechanical Drafting and other technical areas, there is a frequent need to represent sections, cuts, or materials. This is achieved by applying a hatch, composed by a regular pattern, a single color or a gradient between two colors, fulfilling an area. In this chapter we present the related commands.

Topics covered in this chapter:

- Applying hatches composed by families of lines
- Applying fills
- Applying gradients of colors
- Editing applied hatches

Applying patterns

When there is the need to represent a section that includes material cut, the material must be correctly represented by a hatch. Also the representation of floor tiles, bricks, roof tiles, woods or other materials can be indicated by an adequate hatch pattern. Alternatively, it is possible to fulfill an area with a solid single color or a gradient between two colors. The following diagram illustrates the hatches application:

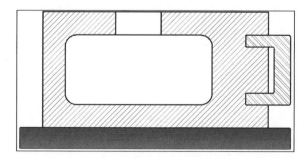

The HATCH command (alias H, 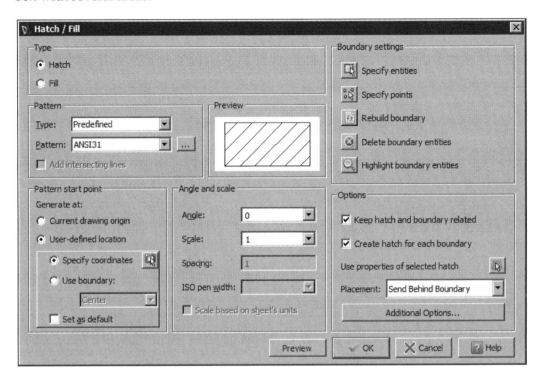 on the **Draw** toolbar, or **Draw | Hatch/Fill** main menu) allows the user to apply hatches. The command displays a dialog box with several areas.

In **Type** it is possible to choose between a **Hatch** (pattern of lines or solid fill) or a **Fill** (gradient, presented later). In **Pattern**, the **Type** list includes three possibilities: **Predefined**, **User-defined** and **Custom**. The **Preview** button allows the user to preview the hatch application and the **OK** button applies it.

The Predefined and Custom types

The **Predefined** type allows the user to select a pattern included with the program, in the **Pattern** list or by pressing the button to the right of the list. This button displays available patterns, divided by categories (**ANSI, ISO, Sample** and **Custom**). In the **Angle and scale** area, it is possible to control the pattern **Angle** and **Scale**. Additionally, just for ISO patterns, the **ISO pen width** can also be specified.

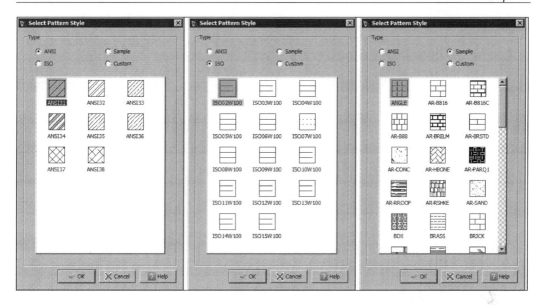

The **Custom** type allows the user to select custom patterns, included in the SAMPLES. PAT file, similar to the **Custom** category in **Predefined**.

 Predefined hatch patterns include the **Solid** pattern, which allows the user to apply solid fills.

User-defined type

The **User-defined** type allows the user to apply a simple pattern, composed by one family of continuous lines or two perpendicular families. With this type, in the **Angle and scale** area, it is possible to control the pattern **Angle** and **Spacing** between lines, in drawing units.

Pattern origin

The **Pattern** start point area allows the user to define the origin reference of the pattern. It can be the **Current drawing origin** or **User-defined location**. This last option allows the user to select one point in the drawing area as origin, use one of the four boundary corners or boundary center. **Set as default** will use the selected option as default for next command applications.

By applying a **User-defined location**, it is easy to adjust the pattern to a boundary corner.

Boundary definition

Besides specifying a pattern and its properties, the boundary must be defined. In the **Boundary settings** area, the **Specify entities** button temporarily hides the box and allows the user to select all entities that limit the area to be hatched. The **Specify points** button also temporarily hides the box and prompts for picking points inside the areas to be hatched. In each pick, the area boundary is highlighted or, if the boundary can not be defined (not closed or outside the drawing area), a message is displayed. The **Rebuild boundary** button is only available for editing hatches (EDITHATCH command, similar to this one) and allows the user to create a boundary for an existing hatch. The **Delete boundary entities** button allows the user to remove entities that are already selected. Finally, **Highlight boundary entities** hides the box just to display current boundaries.

 The most used option to define boundaries is **Specify points**. If the boundary cannot be defined, maybe increasing gap size or drawing temporarily crossing lines will solve the problem.

Other options

The **Hatch/Fill** dialog box includes some additional options in the **Options** area. With **Keep hatch and boundary related** checked, the hatch is associated to the boundary; if this changes, the hatch updates. With **Create hatch for each boundary** checked, even if several areas are defined for hatching with a single command, hatches are separated, allowing for individual deletion or modification. Properties of an existing hatch can be specified by pressing the **Use properties of selected hatch** button. The hatches order, relating to boundaries, are specified in the **Placement** list, normally **Send Behind Boundary**.

The **Additional Options** button displays a new dialog box. The **Internal regions** area controls the command behavior when there are internal regions. If **Find internal regions** is checked, when selecting entities or specifying internal points, the command verifies if there are internal boundaries. If this is the case, there are three possibilities: **Out**, the hatch stops at the first boundary; **In/Out**, the hatch alternates between boundaries; **Ignore** applies the hatch without caring for internal boundaries. **Origin** controls the chosen origin when applying properties of an existing hatch. **Gaps** allows defining a **Maximum gap size** so the hatch is applied even with an open boundary. **Boundary group** indicates whether boundaries are located within the **Active View Tile** or within a selection accessed by the button. Finally, **Boundary preservation** defines whether boundaries are kept in the drawing after applying hatches and which type of entities, if polylines or regions.

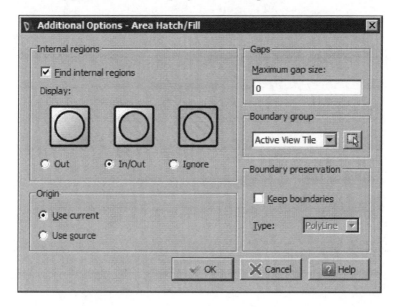

Applying gradients

The HATCH command includes the possibility for filling areas with a gradient. In **Type**, by selecting **Fill**, other options are available, as displayed in the following screenshot:

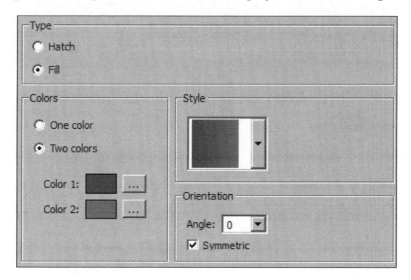

The gradient can be defined between **One color**, black or white (**Dark** and **Light**), or **Two colors**. The **Style** list includes all available gradient styles. **Orientation** allows rotating the gradient fill according to the **Angle** list. With **Symmetric** checked, the gradient is applied symmetrically relating to the boundary.

Editing fills and patterns

The EDITHATCH command (alias HE on the **Modify** toolbar, or by navigating to **Modify | Entity | Hatch**) allows modifying existing hatches and fills. The command displays a dialog box similar to the HATCH command box.

> The only difference is the **Rebuild boundary** option, which allows the user to recreate the boundary of an unbounded hatch.

Exercise 8.1

Continuing the mechanical project from the last chapter, we are now applying hatches to Section AA.

1. Open the drawing PROJECT1-CH7.DWG.

2. Activate the **HATCH1** layer.

3. To hatch the support (larger part) we are going to apply a **User-defined** hatch. Apply the HATCH command, in **Pattern Type** select **User-defined**, in **Angle** select 45 and, in **Spacing**, type 3. Press the **Specify points** button and pick a point inside the bottom part. The boundary is highlighted. Press *Enter* to exit the drawing area and come back to the dialog box. Now, press the **Preview** button to see if the hatch is correctly applied. Click the mouse right-button to accept the preview and end the command.

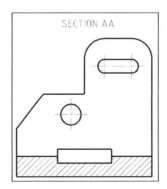

4. Activate the **HATCH2** layer.

5. The pin part will be hatched with an ANSI predefined pattern. Apply the HATCH command again. In **Pattern Type** select **Predefined** and click on the button to the right of the **Pattern** list. In the **Select Pattern Style** box, ANSI Type, select **ANSI32**. Now, in **Angle** select 90 and, in **Scale**, type 0.5. Press the **Specify points** button and pick a point inside the pin part. The boundary is highlighted. Press *Enter* to exit the drawing area and come back to the dialog box. Again, verify if the hatch is correct with the **Preview** button. Click on the mouse right-button to accept the preview and end the command.

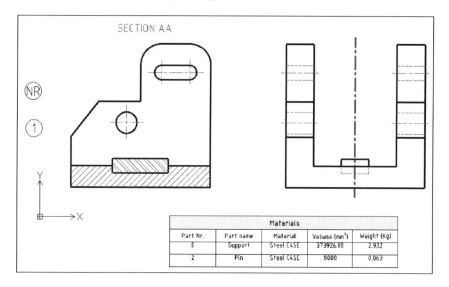

6. Save the drawing with the name PROJECT1-CH8.DWG. In the next chapter we will give dimensions to the drawing.

Summary

In this chapter, hatches are presented, allowing the user to represent sections, cuts, or materials. Hatches can be a regular pattern composed by families of lines, a single color or a gradient between two colors. The HATCH command allows the user to select a pattern or a gradient and apply it to closed or almost closed areas. The EDITHATCH allows the user to modify existing hatches and fills.

The next chapter includes all commands for dimensioning or documenting drawings, another essential step for mechanical drafting.

9
Documenting Projects

Another essential phase of any mechanical project is documenting it. This is done by applying all the necessary dimensions to manufacture the project. This chapter includes the most important and regularly used dimension commands.

In this chapter we will cover the following topics:

- Creating horizontal, vertical, and aligned dimensions
- Creating continued and baseline from existing dimensions
- Creating angular dimensions
- Dimensioning arcs and circles
- Applying ordinate dimensions
- Applying leaders
- Editing and arranging dimensions
- Creating and editing dimension styles

Creating linear dimensions

Normally, a dimension is composed by a dimension line, a text or value, two extension lines, and two arrows. The most common dimensions are linear, presented next.

When creating the first dimension in a drawing, a layer called
Defpoints is added to the drawing. This layer stores the points
that define dimensions and can not be purged, even if it is empty.
However it can be renamed.

Dimensions are placed in the current dimension style. The dimension
style controls all geometric and unit's parameters. The PROPERTIES
palette can be used to apply other dimension style to selected
dimensions.

The LINEARDIMENSION command (alias DLI, ▦ on the **Dimension** toolbar, or
Dimension | Linear main menu) creates horizontal and vertical dimensions.

By default, the command prompts for the first and second points that define what
we want to dimension and a third point that specifies the dimension line location.
The measured value is written on the command line:

```
: LINEARDIMENSION
Default: Entity
Options: Entity or
Specify first extension line position»P1
Specify second extension line position»P2
Options: Angle, Horizontal, Note, Rotated, Text, Vertical or
Specify dimension line position»P3
Dimension Text : 30
```

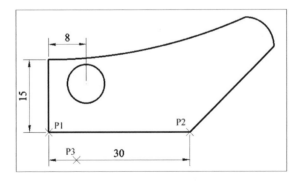

Instead of defining the first extension line position, if pressing the *Enter* key or
applying the **Entity** option, it is possible to select an entity. The **Angle** option allows
the user to rotate the text. The **Horizontal** and **Vertical** options allow forcing a
horizontal or vertical dimensions. The **Note** and **Text** options allow checking and
modifying the text or value. Finally, the **Rotated** option allows the user to rotate the
dimension line related to the extension lines as displayed in the next image;
this option prompts for the dimension line angle.

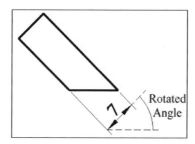

 Instead of applying the **Angle, Note** or **Text** options, it is better to apply the PROPERTIES palette, this allowing editing several dimensions at once.

The PARALLELDIMENSION command (alias DAL, on the **Dimension** toolbar, or **Dimension | Aligned** main menu) creates aligned dimensions.

By default, the command is similar to the previous command, prompting for the first and second points that define what to dimension and a third point that specifies the dimension line location:

```
: PARALLELDIMENSION
Default: Entity
Options: Entity or
Specify first extension line position» P1
Specify second extension line position» P2
Options: Angle, Note, Text or
Specify dimension line position» P3
Dimension Text : 25
```

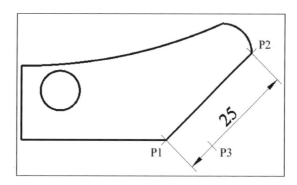

The **Entity, Angle, Note** and **Text** options are similar to the same options of the previous command.

The CONTINUEDIMENSION command (alias DCO, 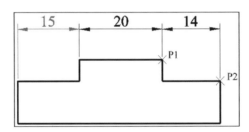 on the **Dimension** toolbar, or **Dimension | Continue** main menu) creates continued dimensions from an existing dimension. By default, the command tries to continue from the previous dimension, or, if not available, prompts for selection. If the previous dimension is not the wanted one, we press *Enter* to select it. Then, it's only to specify second extension lines points, and *Enter* twice to leave the command:

```
: CONTINUEDIMENSION
Specify dimension»Selection
Default: Select dimension
Options: Select dimension, Undo or
Specify second extension line position»P1
Dimension Text : 20
Default: Select dimension
Options: Select dimension, Undo or
Specify second extension line position»P2
Dimension Text : 14
Default: Select dimension
Options: Select dimension, Undo or
Specify second extension line position»Enter
Specify dimension» Enter
```

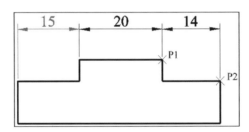

The **Undo** option allows the user to undo the last continued dimension. When pressing *Enter* the first time, it is possible to select another dimension and continue from that one.

The BASELINEDIMENSION command (alias DBA, on the **Dimension** toolbar, or **Dimension | Baseline** main menu) creates parallel dimensions, sharing a common extension line, from an existing dimension. By default, the command tries to apply baseline from the previous dimension, or, if not available, prompts for selection. If the previous dimension is not the wanted one, we press *Enter* to select it. Then, it's only to specify second extension lines points, and *Enter* twice to leave the command:

```
: BASELINEDIMENSION
Specify base dimension»Selection
Default: Base dimension
Options: Base dimension, Undo or
Specify second extension line position» P1
Dimension Text : 35
Default: Base dimension
Options: Base dimension, Undo or
Specify second extension line position» P2
Dimension Text : 49
Default: Base dimension
Options: Base dimension, Undo or
Specify second extension line position» Enter
Specify base dimension» Enter
```

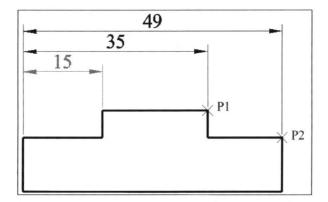

The **Undo** option allows the user to undo the last baseline dimension. When pressing *Enter* the first time, it is possible to select another dimension and apply baseline from that one.

 The distance between dimension lines is defined by the dimension style: **Line** | **Dimension line settings** | **Offset**.

Creating angular dimensions

The ANGLEDIMENSION command (alias DAN, 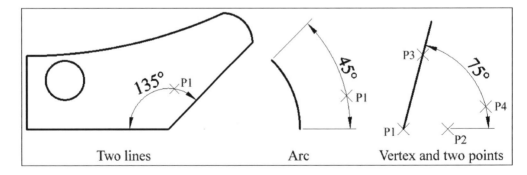 on the **Dimension** toolbar, or
Dimension | Angular main menu) creates angular dimensions. By default,
it prompts for an entity selection. If the entity is a line, it prompts for the second
line. Then, it prompts for dimension line location:

```
: ANGLEDIMENSION
Options: Enter to specify vertex or
Specify entity» Selection
Specify second line» Selection
Options: Angle, Note, Text or
Specify dimension position»P1
Dimension text : 135
```

Instead of selecting an entity, by pressing *Enter* the angle to dimension is defined
by the vertex and two points:

```
: ANGLEDIMENSION
Options: Enter to specify vertex or
Specify entity» Enter
Specify vertex position» P1
Specify angle start point» P2
Specify angle end point» P3
Options: Angle, Note, Text or
Specify dimension position» P4
Dimension text : 75
```

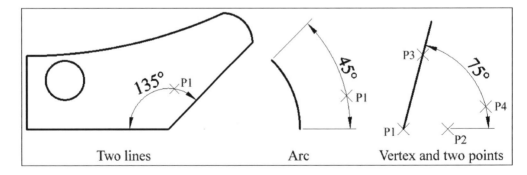

| Two lines | Arc | Vertex and two points |

The **Angle**, **Note**, and **Text** options are similar to the same options of the previous commands.

> The angular unit controls the symbol automatically applied with the default value.
>
> Specifying vertex and two points is the only method to dimension angles greater than 180 degrees.
>
> The command can also be applied to dimension part of a circle, which is almost useless. The selection point defines the first extension line, and a second point is requested.

Dimensioning arcs and circles

This section discusses other commands for creating dimensions.

The RADIUSDIMENSION command (alias DRA, on the **Dimension** toolbar, or **Dimension | Radius** main menu) creates radial dimensions. The command prompts for the selection of an arc or circle and the dimension line location:

```
: RADIUSDIMENSION
Specify curved entity» Selection
Options: Angle, Note, Text or
Specify dimension position» P1
Dimension Text : 4
```

The **Angle**, **Note** and **Text** options are similar to the same options of the previous commands.

> By accepting the default value, the letter R is automatically used as prefix.

The DIAMETERDIMENSION command (alias DDI, on the **Dimension** toolbar, or **Dimension | Diameter** main menu) creates diameter dimensions. The command prompts for the selection of an arc or circle and the dimension line location:

```
: DIAMETERDIMENSION
Specify curved entity» Selection
Options: Angle, Note, Text or
Specify dimension position» P1
Dimension Text : 8
```

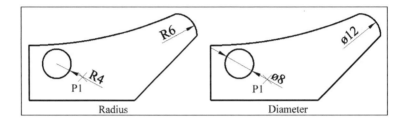

Radius   Diameter

The **Angle**, **Note** and **Text** options are similar to the same options of the previous commands.

> By accepting the default value, the diameter symbol is automatically used as prefix.

The ARCLENGTHDIMENSION command (alias DAR, 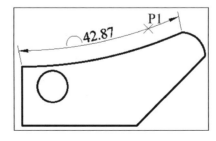 on the **Dimension** toolbar, or **Dimension | Arc Length** main menu) dimensions the perimeter of arcs. The command just prompts for the selection of an arc the dimension line location:

```
: ARCLENGTHDIMENSION
Specify curved entity» Selection
Options: Angle, Note, Partial, Text or
Specify dimension position» P1
Dimension Text : 42.87
```

The **Angle**, **Note** and **Text** options are similar to the same options of the previous commands. The **Partial** option allows the user to dimension only part of the arc.

 By accepting the default value, an arc symbol is automatically used as prefix.

Ordinate dimensions and leaders

Ordinate dimensions correspond to X-Datum or Y-Datum point coordinates. This type of dimensioning is useful, not only for elevations, but also for perforated plates or other areas with a reference point. Before applying ordinate dimensions, it is often necessary to specify a new coordinate system just for indicating a new origin.

The CCS command (also UCS, ▟ on the **CCS** toolbar, or **Tools | New CCS** main menu) creates custom coordinate systems. By default, the command prompts for the new origin, a second point that indicates X direction and a third point that indicates the positive XY plane. If *Enter* is pressed when prompted for the second point, a new custom coordinate system is created just by moving origin:

```
: CCS
Default: World
Options: align to Entity, NAmed, Previous, View, World, X, Y, Z, ZAxis or
Specify origin» P1
Options: Enter to accept or
Specify X-axis through point» Enter
```

The **World** option allows the user to return to the world coordinate system (WCS), the main DraftSight coordinate system.

The ORDINATEDIMENSION command (alias DOR, 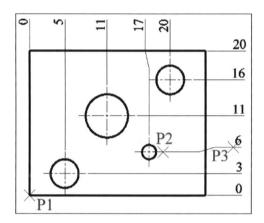 on the **Dimension** toolbar, or **Dimension | Ordinate** main menu) creates ordinate dimensions. By default, it just prompts for the point to be dimension (datum position) and the dimension position:

```
: ORDINATEDIMENSION
Specify datum position»P2
Options: Angle, Note, Text, X datum, Y datum, set Zero or
Specify dimension position»P3
Dimension Text : 6
```

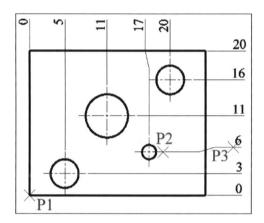

The **Angle**, **Note** and **Text** options are similar to the same options of the previous commands. **X datum** and **Y datum** options allow blocking which direction is dimensioned. **set Zero** allows the user to create a zero datum and use **Reference** dimensions according to the zero datum previously created.

The SMARTLEADER command (alias LE, 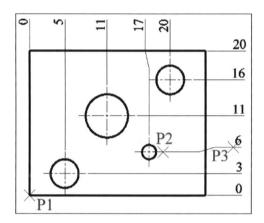 on the **Dimension** toolbar, or **Dimension | Leader** main menu) creates leader dimensions. A leader allows the user to connect a text or block to a special feature of the project. By default, it prompts for the leader tip (where the arrow should be), the end point of arrow and one or more lines of text. Pressing *Enter* once changes line and pressing *Enter* twice ends the command and draws the leader:

```
: SMARTLEADER
Default: Settings
Options: Settings or
```

```
Specify start point» P1
Options: Enter to exit or
Specify next vertex» P2
Options: Enter to exit or
Specify next vertex» Enter
Default: Editor
Options: Editor or
Specify text» 16 holes
Options: Enter to exit or
Specify text» %%c10
Options: Enter to exit or
Specify text» Enter
```

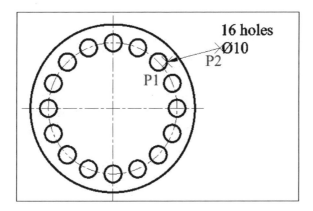

The **Settings** option displays a dialog box allowing configuring the leader. The **Editor** option enters the NOTE command allowing entering text with the multiline text editor.

 %%c is the code for ⌀ (diameter symbol) and %%d the code for ° (degree symbol).

Editing dimensions

Frequently, the default position for dimensions is not adequate. Several methods are available for editing dimensions.

To control wrong extension line points, text, and dimension line positions, the best method is to select the dimension, without command, and move grips. In the following diagram, the horizontal dimension is wrong due to an incorrect extension origin. We select the dimension, activate the grip (turns red) and move it to the correct position.

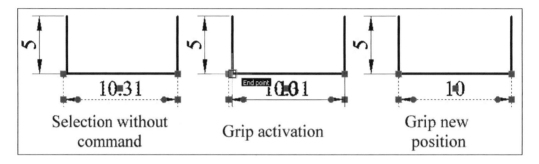

The PROPERTIES palette is excellent for editing the geometric properties of some dimensions and controlling text contents and text angle.

Just to edit text contents, we can also apply the EDITANNOTATION command.

The EDITDIMENSION command (alias DED or ![icon] on the **Dimension** toolbar) includes five options for editing dimensions. The command starts by displaying these options:

```
: EDITDIMENSION
Default: Home
Options: Angle, Home, Move, New, Oblique or
Specify option» Option
```

The **Angle** option allows the user to modify the text angle for the dimensions to be selected. The **Home** option places the text of dimensions to be selected at the default position. The **Move** option allows the user to move the text of one dimension. The **New** option allows the user to type new text for the dimensions to be selected. The **Oblique** option allows the user to skew the extension lines, relating to the dimension line; it prompts for the angle and the dimensions to oblique.

Creating and managing dimension styles

Dimensions are created with the current dimension style. The PROPERTIES palette allows the user to modify the dimension style of selected dimensions.

The DIMENSIONSTYLE command (alias D, 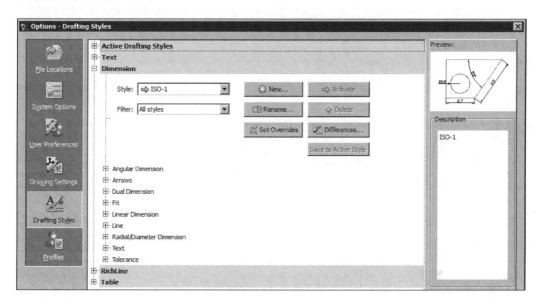 on the **Dimension** toolbar, or **Format | Dimension Style** main menu) allows the user to create and managing dimension styles.

Actually, the command opens the **Options** dialog box, **Drafting Styles** folder, **Dimension** area:

The **Style** list includes all available dimension styles in the drawing. The **New...** button allows the user to create a new style, **Activate** makes current the style in the list, **Rename...** allows the user to rename the style, and **Delete** removes it, if not used. **Set Overrides** allows the user to impose properties to the style and **Differences...** allows the user to compare two styles. Override changes can be inserted in the current style with **Save to Active Style**.

Dimension style settings are organized by expandable tree:

- **Angular Dimension**: This section includes all parameters related to angular dimensions, namely format and precision, zero suppressions and arc length symbol.

- **Arrows**: This section includes all parameters related to arrows, namely start arrow, end arrow, leader arrow and size.

- **Dual Dimension**: This section includes parameters allowing linear dimensions to display values in two units, for instance, millimeters and inches.

- **Fit**: This section includes all parameters related to arrange dimension elements when there is not enough space for text and arrows, and a scale factor that can be applied to all geometric parameters.

- **Linear Dimension**: This section includes all parameters related to linear dimensions, namely format and precision, decimal separator, prefix, suffix, zero suppressions and measurement scale factor.

- **Line**: This section includes all parameters related to dimension lines and extension lines, as suppression, colors and line width, fixed length for extension lines, baseline dimension line offset.

- **Radial/Diameter Dimension**: This section includes parameters related to center mark display (CENTERMARK command) and radius dimension jog (JOGGEDDIMENSION command).

- **Text**: This section includes all parameters related to text, namely text style, color and height, text position related to dimension line and text alignment.

- **Tolerance**: This section includes parameters allowing dimensional tolerances, like maximum and minimum permissible values.

Exercise 9.1

We are now documenting the mechanical project from the last chapter.

1. Open the drawing PROJECT1-CH8.DWG.

2. Activate the **Dimensions** layer.

3. Applying the LINEARDIMENSION command (alias DLI), create the 100 dimension below the front view, marking the left and right endpoints and position the dimension line.

4. Applying the same command, create all dimensions displayed in the following image. If necessary, move the text up a bit and, with grips, increase the length of the cutting plane line.

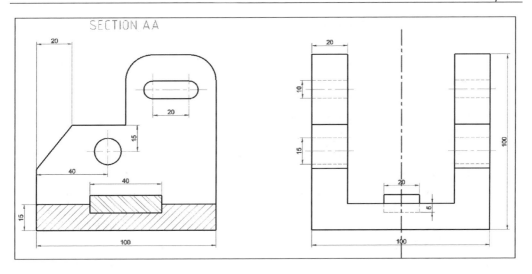

5. Now, applying the CONTINUEDIMENSION command (alias DCO) complete all linear dimensions displayed in the following image. With a single command it is possible to create all continued dimensions. Pressing *Enter* once, it is possible to continue another dimension, pressing *Enter* twice or *Esc* key ends the command.

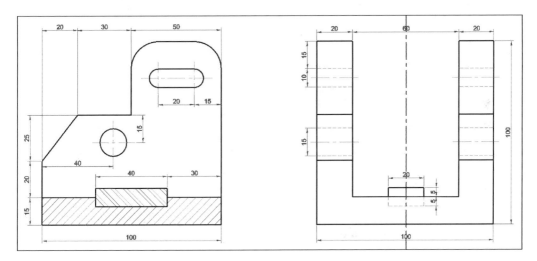

6. The last dimension missing is the radius of the top fillet, done with the RADIUSDIMENSION command (alias DRA).

7. The 15 dimension on the right, corresponding to the circular hole must include a diameter symbol. Clicking twice on that dimension opens the PROPERTIES palette and, in **Text override** parameter, type %%C15.

8. The two 5 vertical dimensions on the right don't look right. Again with the PROPERTIES palette, replace the second arrow of the bottom dimension and the first arrow of the top dimension with a dot.

9. The text of horizontal dimensions intersected by the cutting plane line must be pushed to the right. This is done by selecting each dimension and, with grips, activating the text grip and move it.

10. We are now going to apply leaders with the block already created. Apply the SMARTLEADER command (alias LE) and **Settings** option to format the leader. In the **Format Leaders** dialog box specify **Type Block (Annotations** tab) and **Arrow style Dot (Arrows/Lines** tab). Confirm by clicking on **OK**. Mark the start point inside part 1, second point outside, and *Enter* for ending points. Type the name block NUMBER and accept scales and rotation. When prompting for **Part number**, type 1. Insert another leader for part 2, but now typing 2 as part number.

11. The block insertion point is at the circle center. Apply the EDITCOMPONENT command (for instance, select a block and right button mouse menu) and move the base point to the left quadrant. Exit the block edition, saving changes.

12. Activate the **Cutting Text** layer.

13. Apply again the SMARTLEADER command (alias LE) for specifying section letters. With **Settings** option, select **Note** as **Type** and the bottom left justification (text over line), and, in **Arrows/Lines** tab, specify **First segmentHorizontal** and **Arrow style Closed filled**. Draw the top arrow and type A as text. Copy this arrow to the other endpoint of the cutting plane line.

14. Erase the isolated block insertion and original block entities, coming from *Chapter 7, Creating and Applying Components*. The project should be similar to the following screenshot.

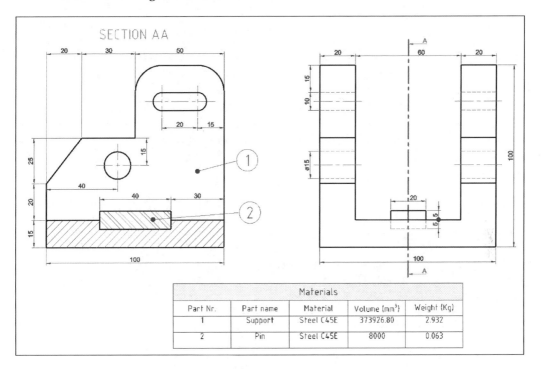

Materials				
Part Nr.	Part name	Material	Volume (mm³)	Weight (Kg)
1	Support	Steel C45E	373926.80	2.932
2	Pin	Steel C45E	8000	0.063

15. Save the drawing with the name PROJECT1-CH9.DWG. In the next chapter we will prepare the drawing to print.

Summary

This chapter includes the commands used to document or dimension a drawing. The LINEARDIMENSIONcommand creates horizontal and vertical dimensions. The PARALLELDIMENSION command creates aligned dimensions. The CONTINUEDIMENSION and BASELINEDIMENSION commands create continued or parallel dimensions from an existing dimension. The ANGLEDIMENSION command creates angular dimensions. The RADIUSDIMENSION and DIAMETERDIMENSION commands create radial and diameter dimensions and the ARCLENGTHDIMENSION command dimensions the perimeter of arcs. The ORDINATEDIMENSION command creates ordinate dimensions, corresponding to X-Datum or Y-Datum point coordinates. Before applying ordinate dimensions, it is often necessary to specify a new coordinate system to define the origin, which is done by the CCS command. The SMARTLEADER creates leader dimensions, connecting a text or block to a special feature of the project.

There are several methods for editing dimensions: selecting without commands and use grip editing for placing dimension elements, applying the PROPERTIES palette, or applying the EDITDIMENSION command.

The DIMENSIONSTYLE command allows the user to create and manage dimension styles, which controls the aspect and measurement of dimensions, namely geometric parameters and units.

In the next chapter we will prepare sheets (layouts) for printing at precise scales.

10
Printing Efficiently

When completing a project, a common task is preparing it for printing. This chapter instructs how to prepare a drawing for printing, namely configuring sheets, creating viewports, adjusting their scales, previewing and printing.

Topics covered in this chapter are as follows:

- Creating, renaming, and deleting sheets
- Configuring sheets
- Creating and managing print styles
- Creating viewports
- Adjusting viewports' visualization
- Adjusting viewports' scales
- Controlling layer properties per viewport
- Previewing
- Printing sheets

Creating and managing sheets for printing

There are some tabs present below the drawing area. The first, named **Model**, is the one used for modeling and creating drawings. The other tabs, by default named **Sheet1** and so on, represent simulations of paper sheets that will be printed. It is possible to define up to 255 sheets, also known as **layouts**.

 DraftSight also allows the user to print from the model, but printing from sheets is much more versatile.

Printing from sheets has the following advantages:

- Sheets can include several viewports, each with its own view and scale. So, to add details, it's only to create a new viewport zoomed to the part and with a different scale

- From a single model, it is possible to define several sheets, aiming different printers, paper formats, or with distinct information

- Elements like annotations, title blocks, identification, notes, tables, and so on, can be directly inserted in sheets, without mixing it with the model

- Layer properties can be controlled by viewport, like freezing, modifying colors, linestyles and line widths

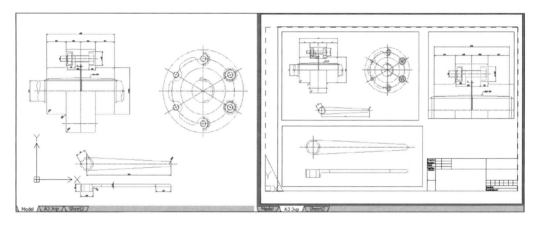

The easiest way to manage sheets is by clicking the right-mouse button over these tabs, as displayed in the following image:

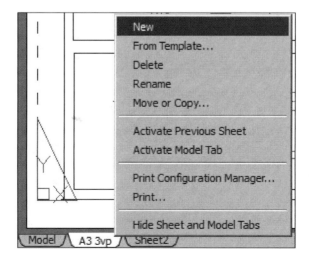

New allows the user to create a new sheet and **From Template** allows the user to copy a sheet from a template file. The new sheet is given the name **Sheet** with the lowest number available. **Delete** and **Rename** allows the user to erase or rename sheets. The **Model** tab cannot be erased or renamed. With **Move or Copy** it is possible to reorder or copy sheets.

Additional options in this menu allow the user to activate the model tab or the last sheet tab, accessing the PRINTCONFIGURATION and PRINT commands and hiding tabs.

The SHEET command (**File | Sheet** main menu) also includes these options.

[A better way to rename sheets is by double-clicking on the tab and typing the new name.]

Configuring sheets

After creating and activating a sheet, this must be configured with the next command.

The PRINTCONFIGURATION command (click on the mouse's right-button over the tab, or **File | Print Configuration** main menu) allows the user to configure sheets. It starts by displaying the **Print Configuration Manager** dialog box, including print configurations associated with the drawing.

We can create print configurations and apply them to the model or any sheet in any drawing. The **New** button creates print configurations, based on a **Default** or existing configurations, and displaying a file dialog box to save the .CFG file. By default, this file is saved in the **Print Settings** folder associated to the DraftSight user profile. After saving it, the **Print Configuration** dialog box is displayed. The **Edit** button allows the user to edit the selected configuration, also displaying the **Print Configuration** box. The **Import** button allows the user to import configurations to the current drawing by selecting a .CFG file or included in another drawing file. The **Activate** button allows the user to select an existing configuration for the current sheet. The **Assign to** button allows the user to assign the selected configuration to one or more sheets of the drawing. With **Show dialog box on creation of new sheets** checked, this box is displayed every time a new sheet is created and accessed.

The **Print Configuration** dialog box, presented in the next image, controls all configuration parameters and is similar to the **Print** dialog box, presented later in this chapter.

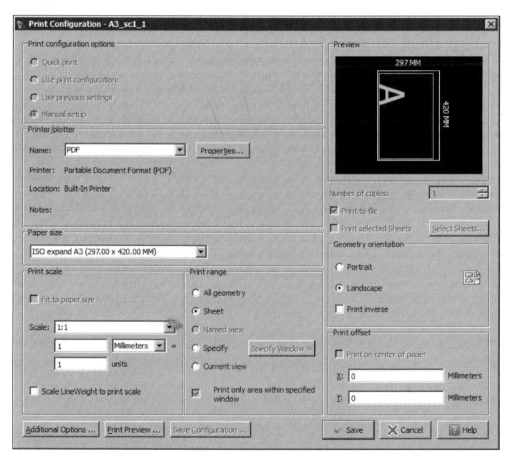

This box includes the following areas:

- **Print configuration options**: These options are only available in the **Print** dialog box, thus explained later.

- **Printer/plotter**: The printer or printer driver is selected in the **Name** list. Besides all printers or plotters available in the operative system, there are some DraftSight printer drivers that create files, namely PDF, JPG, PNG, and SVG (scalable vector graphics). Depending on the choice, the **Properties** button displays a dialog box with corresponding properties. To all printer drivers coming with the program, it is only possible to control paper's printable area.

- **Paper size**: Depending on the chosen printer or printer driver, in this list we select the paper size.

- **Print scale**: Here the paper scale is defined. It can be **Fit to paper size**, to get the maximum paper area, a standard scale (in the **Scale** list) or a user-defined scale. The units list only considers **Millimeters** or **Inches**, but it is possible to define a ratio between printed units and drawing units. **Scale LineWeight** to print scale allows the user to adjust lineweights.

If we are simulating the paper to be printed, the 1:1 ratio must be used, and viewport scales are defined with the PROPERTIES palette. But if our drawing is not in millimeters or inches, we must "tell" the program that unit. This is easily done by placing our unit value in millimeters or inches. For instance, if our drawing is in meters, we type 1000 millimeters to 1 unit.

Scale Line Weight option must normally be unchecked, so that the lineweights are not modified with print scale.

- **Print range**: This area includes several options for defining what to print. **All geometry** (bounding box of visible entities), the **Sheet**, a **Named** view, part of the drawing selected by the **Specify Window** button, or the **Current view**.

Again, if we are simulating the paper to be printed, the **Sheet** option is the correct one.

- **Number of copies, Print to file, Print selected Sheets**: These options are only available in the **Print** dialog box, thus explained later.

- **Geometry orientation**: Here the orientation of the drawing is defined, between **Portrait** and **Landscape**. **Print inverse** applies a 180 degree rotation. The **Preview** area shows the paper size and area to be printed.

- **Print offset**: It is possible to adjust the area of printout, by centering (not available with **Sheet**), or **X** or **Y** axis margins.

The **Additional Options** button displays a new dialog box with three areas:

- **Options**: This area includes options to apply hiding geometry (if the HIDE command is used in viewports, only for 3D models), if sheet entities are sent to the front related to model entities, if entities lineweights are used (instead of print styles), or if the assigned print style is used.

- **Shaded views**: These options control viewport quality with shaded views (3D models).

- **PrintStyle table**: The print style, which defines, among other options, colors and lineweights, is selected in the list. The **New** button allows the user to create a new style and the **Edit** button allows the user to edit the selected style. Print styles are explained next.

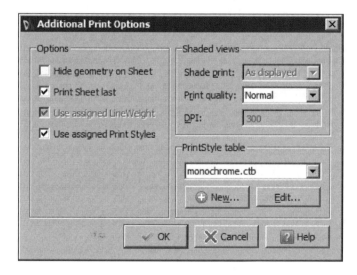

Creating and managing print styles

Print styles control how entities are printed. The most important properties to control are printed colors and lineweights. Normally, these properties are based on entities or layers colors, thus being defined by CTB files. Print style files are saved in the `Print Styles` folder associated to the user profile.

When editing a CTB file, for instance from the **Print Configuration** dialog box, the **Print Style Table** dialog box is displayed. The principal area of this box is **Formats**. Here, for each main color (1 to 255) it is possible to assign several printing properties, the most important being **LineColor** (printed color) and **LineWeight** (printed lineweights).

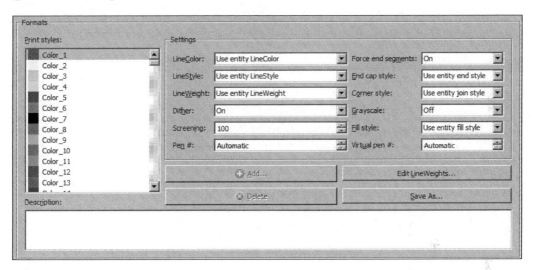

The same print style file can be applied to more than one sheet or to any drawing.

Actually, there are two types of print styles: the CTB files, where printouts are based on entities or layers colors, and the STB files, where print style names are entities or layers properties. A drawing, when created, is assigned to use CTB or STB print styles. The CONVERTPRINTSTYLES command allows the user to convert a drawing from STB to CTB, or CTB to STB.

Creating viewports

After defining a sheet, one or more viewports must be created. A viewport is like a window that displays our drawing, or part of it. By default, when accessing a sheet for the first time, a viewport is created. As this sheet was not configured, the viewport may be deleted.

The VIEWPORT command (alias MV, **View | View Tiles** main menu) creates viewports in sheets. By default, it prompts for two opposite corners and creates a rectangular viewport.

```
: VIEWPORT
Default: Fit
Options: 2, 3, 4, Align, Entity, Fit, Lock, OFf, ON, Polygonal, Restore,
SHaded view or
Specify start corner»P1
Specify second corner»P2
```

The **2**, **3**, and **4** options allow creating 2, 3, or 4 viewports between two opposite corners and respective distribution. The **Align** option allows the user to align existing viewports. The **Entity** option prompts for a closed entity selection and applies this entity as a viewport boundary. The **Fit** option creates a single viewport fitting the paper (without printer margins). The **Lock** option allows the user to lock viewport's view, so zooming inside does not affect scale. **OFF** turns off the viewport view and **ON** turns it on. The **Restore** option allows the user to restore a saved viewport configuration. The **Shaded view** option controls the representation of shaded views.

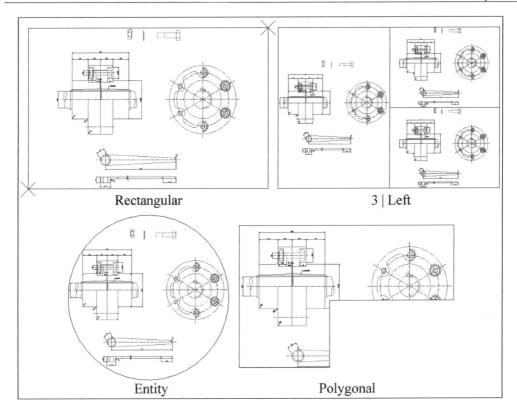

Rectangular   3 | Left

Entity   Polygonal

Adjusting viewport visualization, scale, and layers

When viewports are created, by default the program applies a zoom extents displaying inside all the drawings.

To adjust visualization inside each viewport, we must access the model inside the viewport. Then, we may want to modify other viewports' visualization and, after, come back to sheet space. So, there are three possibilities for navigating between viewports and spaces:

- **Double-clicking inside a viewport**: We access the model inside that viewport, so we can edit the model, but, most important, we can adjust visualization and layer properties. The active viewport has a thicker border and the cursor is normal inside, when outside is displayed as an arrow. This is the same as applying the MODELMODE command (alias MM or MS).

- **Single-clicking inside another viewport**: If one viewport is active, a single-click inside another viewport activates it.

- **Double-clicking anywhere outside viewports**: To come back to sheet mode, we double-click outside viewports. The model cannot be edited and the cursor is normal anywhere. This is the same as applying the SHEETMODE command (alias SM or PS).

In almost all technical areas, printing the drawings at precise scales is important. To adjust viewports' scales, in DraftSight, the easiest way is to select each viewport and apply the PROPERTIES palette.

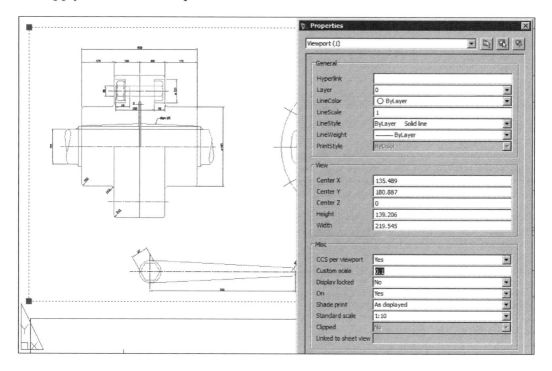

In **Standard scale**, it is possible to select a predefined scale. If we want to apply other scales, this can be done in **Custom scale**.

 There is no need to make the calculation for a **Custom scale**. For instance, if we want to apply a 1 : 15 scale, not predefined, it is enough to type 1/15.

Some layer properties can be controlled per viewport. With an active viewport, the **Layers Manager** dialog box (LAYER command, alias LA) includes the following additional columns that allow for modifying properties only for that viewport: **Active ViewPort** to freeze layers, **VP Color** to modify layer colors, **VP LineStyle** to modify layer linestyles, **VP LineWeight** to modify layer lineweights, and **VP PrintStyle** to modify layer print styles (if STB print styles are used).

Previewing and printing

The PREVIEW command (alias PRE, 🔍 on the **Standard** toolbar, or **File** main menu) displays a print preview of the current sheet in a new box. We can zoom and pan the preview, to see if everything is OK before printing.

The PRINT command (shortcut *Ctrl + P*, 🖨 on the **Standard** toolbar, or **File** main menu) prints the sheet. It displays the **Print** dialog box, similar to the **Print Configuration** box, but allowing access to some specific areas:

- **Print configuration options**: The **Quick print** option prints the screen view with default printer and other default settings. The **Use print configuration** allows the user to select another print configuration. The **Use previous settings** option applies the same settings as in the last print. **Manual setup** allows the user to configure everything.

- **Number of copies**: If a printer is selected, it is possible to print several copies.

- **Print to file**: If a printer is selected, this option allows the user to create a file, instead.

- **Print selected Sheets**: This option allows the user to print several sheets at once, selected with the **Select Sheets** button.

- **Save Configuration**: This button allows the user to create a print configuration from current settings.

Exercise 10.1

Following the mechanical project, a sheet will be configured with three viewports. Files for this exercise are available on the book's page, at www.packtpub.com.

1. Open the drawing PROJECT1-CH9.DWG.

2. Double-click on the **Sheet1** tab and type the name A3_PDF.

3. If a viewport is automatically created, erase it.

4.   Before configuring the sheet, if this drawing uses STB files, we need to convert it to CTB, by applying the CONVERTPRINTSTYLES command.

5.   We are going to configure the sheet. With the mouse right-button menu over the **A3_PDF** tab, select **Print Configuration Manager**. Click on the **New** button, confirm **Default** as the base configuration and create a configuration file named A3&PDF.

6.   In the **Print configuration** box, select the **PDF** printer driver and the **ISO expand A3 (297.00 x 420.00 MM)** paper size. Accept the 1:1 scale and Sheet print range. In **Geometry orientation**, select **Landscape**. Clicking on the **Additional Options** button, select the **monochrome.CTB** print style. Click on the **OK** and **Save** buttons. In the **Print Configuration Manager** box select **A3&PDF**, click on the **Activate** button and **Close**.

7.   Apply the LAYER command to create a layer called **VIEWPORTS**, color **White, LineWeight 0.13 mm** and property **NoPlot**. Activate the layer.

8.   With the VIEWPORT command, create two rectangular viewports, as displayed in the following screenshot:

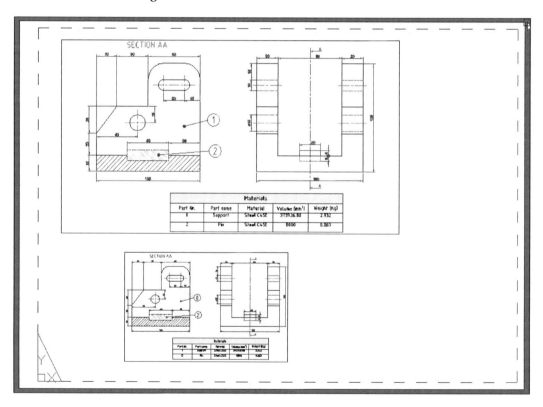

9. With a double-click, enter inside the lower viewport and zoom in to part 2 (smaller part) on the right view. Double-click outside viewports to return to sheet mode.

10. Scales are defined with the PROPERTIES palette (*Ctrl + 1*). Set 1:1 scale to the upper viewport and 2:1 scale to the lower viewport.

11. With grips, adjust viewports sizes, so nothing important is cut.

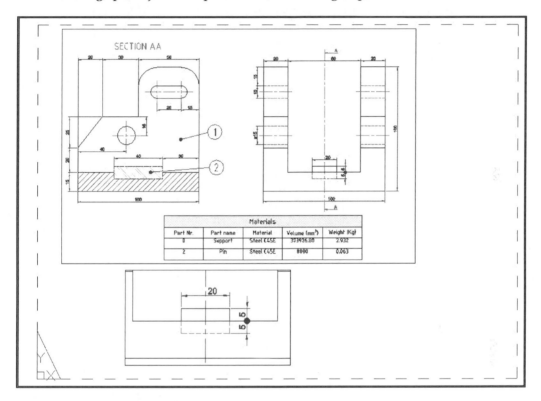

12. Create and activate a layer called **FRAME**.

13. We may insert a title block. Apply the INSERTBLOCK command, click on the **Browse** button and select the titleblock-A3.DWG file. In **Position**, specify the insertion point -16 in X and -6 in Y. These values correspond to the margins of this format paper to the PDF printer driver.

> In sheets, the origin is the lower left corner of the available area, not the paper corner.

14. Activate the **TABLE** layer.

15. Apply the SIMPLENOTE command to type Scale 2:1 near the lower viewport, with 4 mm height.

16. As we are going to create a monochrome PDF, the yellow background on the title table will be black. The PREVIEW command will show this. So, apply the TABLESTYLE command, in **Contents** select **Title** and, in **Background color**, select **None**.

17. Apply the PREVIEW command to verify if everything is OK.

18. Finally, apply the PRINT command to create the PDF file. Give the name PROJECT1.

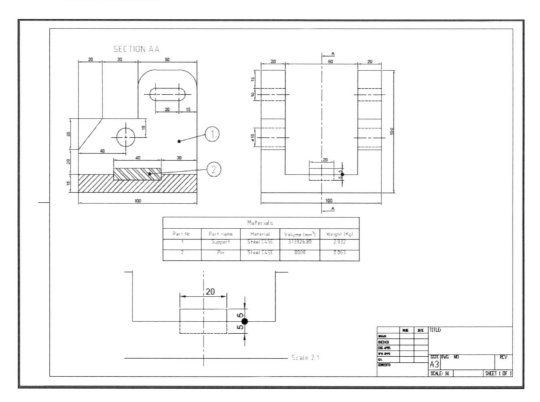

19. Save the drawing with the name PROJECT1-CH10.DWG. In the next chapter we will apply some advanced layer commands.

Summary

This chapter includes the preparation of sheets for printing. The concept of sheet was introduced as well as its advantages. The easiest way to create, rename, or delete sheets is by applying the mouse right-button over the tabs below the drawing area. The PRINTCONFIGURATION allows the user to configure sheets, including the selection of printer/printer driver, paper size, printing scale, what to print, print styles (namely colors and lineweights), and other settings. Sheets are printed 1:1, according to our model unit, as scales are set by viewport.

After sheet configuration, viewports are created with the VIEWPORT command. Double-clicking inside a viewport, we access the model inside to adjust visualization and layer properties. A single-click allows the user to select other viewport and double-click outside viewports comes back to sheet mode. To define viewports' scales we select each viewport and apply the PROPERTIES palette.

The PREVIEW command displays a print preview of the current sheet. The PRINT command prints the sheet, displaying a dialog box, similar to the **Print Configuration** box.

In the next chapter, the last one, we will present some advanced concepts and commands, namely linking other drawings or images and advanced layer commands.

11

Advanced Tools

In this last chapter, some advanced concepts and commands are introduced, namely referencing other drawings and images, and additional layer of commands which are very useful when dealing with complex drawings.

Topics covered in this chapter are as follows:

- Managing other drawings and images linked to our drawing
- Referencing drawings
- Referencing images
- Clipping referenced drawings and images
- Applying advanced layer operations

Referencing external drawings and images

DraftSight allows referencing of other drawings that is, linking other DWG files to our drawing. With this, we can view and print connected information without adding it permanently to our drawing. There are two great advantages: smaller drawing files (as part of the information is outside), and immediate updates. When opening a drawing with references, these are loaded at the current state and cannot be directly modified.

With images it is similar. We can reference, view, and print images, but these are never imported into our drawing.

Managing and linking external drawings and images

The REFERENCES command (alias ER 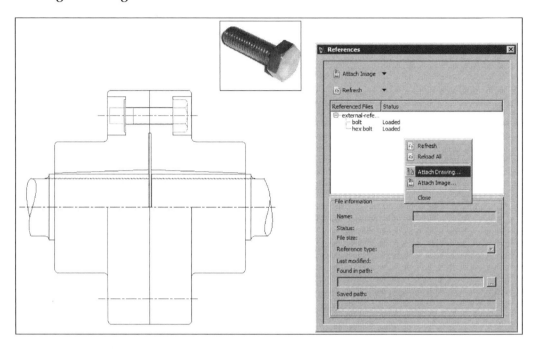 on the **References** toolbar, or **Tools | References Manager** main menu) displays a palette to manage referenced external drawings and images.

The **Referenced Files** area lists all references in the current drawing and their status. When selecting an item, its information is listed in the **File information** area.

The second list includes the following options: **Refresh** to update the list, and **Reload All** to reload all referenced files with last saved changes.

To attach a drawing, we may access the top list or click on **Attach Drawing** by right-clicking on the mouse. Another possibility is to directly apply the ATTACHDRAWING command (alias XA or on the **References** toolbar).

A file dialog box is displayed to select the DWG file, after which the **Attach Reference: Drawing** dialog box is displayed.

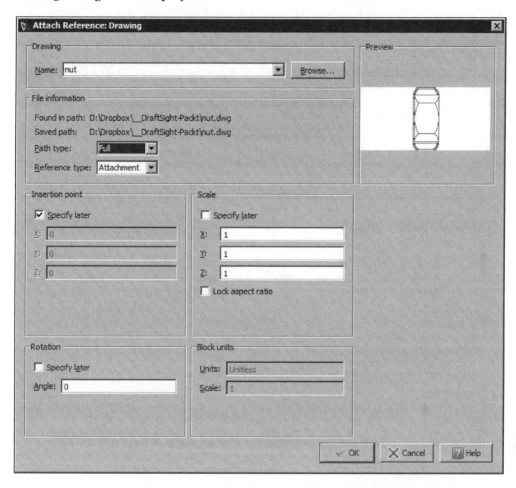

Browse allows 'the user to select another DWG file. **File information** displays the found and saved paths information and two additional lists: **Path type** controls if the path is fully saved (**Full**), not saved (**None**), or saved but without disk letter (**Relative**); **Reference type** controls if the referenced drawing is included (**Attachment**), or not (**Overlay**), if our drawing will be attached to another drawing. The remaining options are similar to a block insertion, namely **Insertion point**, **Scale** along all directions and **Rotation**.

In the **References** box, when selecting a referenced drawing and clicking on the right mouse button, additional options are available: **Open** allows the user to open the referenced drawing (as like applying the OPEN command); **Attach** allows the user to attach the drawing in another position; **Unload** temporarily uploads the drawing and **Reload** reloads it, in its most recent version; **Detach** detaches the drawing and erases all insertions; **Bind** allows the user to transform the referenced drawing into a block, also cutting connection to original file.

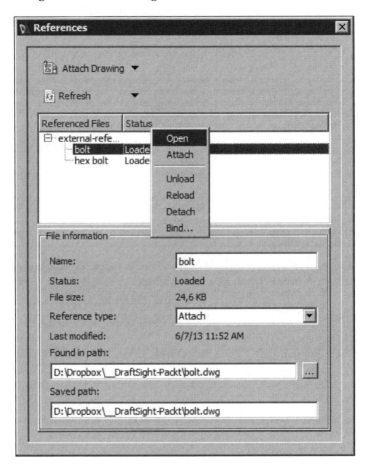

 Symbol names from referenced drawings, as blocks and layers, include the drawing name, a vertical bar, and the symbol name, such as BOLT | LAYER1.

To attach an image, we may access the top list or apply the right mouse button menu. Another possibility is to directly apply the ATTACHIMAGE command (alias IAT or on the **References** toolbar). DraftSight supports the following image file types: BMP, GIF, JPG, JPEG, PNG, TIF, and TIFF.

After selecting the image file, the **Attach Reference: Image** dialog box is displayed. Options are similar to the **Attach Reference: Drawing** dialog box, except with the omission of reference type. Images do not have the **Bind** option.

> If we don't want to see and print the entire external drawing or image, it is possible to clip it with the upcoming commands.

Clipping external drawings and images

The CLIPREFERENCE command (alias XC 🔲 on the **References** toolbar, or **Modify | Clip | Reference** main menu) allows the user to clip referenced drawings or blocks. By default, the command prompts for the selection of one or more references or blocks, a boundary definition, rectangular, and the two corners of the rectangular boundary.

```
: CLIPREFERENCE
Specify reference(s)» Selection
Default: create Boundary
Options: ON, OFf, Delete, create Boundary, Clip depth or Polyline
Specify clipping option» Enter
Default: Rectangular
Options: Invert clip, Polygonal, Rectangular or Select polyline
Specify boundary option» Enter
Specify start corner»P1
Specify opposite corner»P2
```

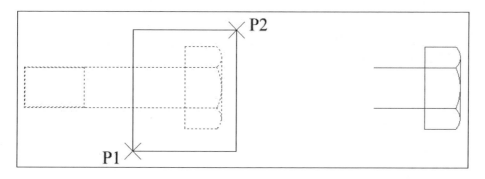

The **ON** and **OFF** options allow the user to turn clipping on and off. **Delete** erases the boundary. **Clip depth** allows the user to define parallel horizontal planes for 3D references. **Polyline** creates a polyline coincident with the existing boundary. The **Create Boundary** option allows the user to define a boundary or modify an existing one. Further options are: **Invert clip** to display only outside the boundary, **Polygonal** to define a polygonal boundary by several points, and **Select polyline** to select an existing polyline as the boundary.

The CLIPIMAGE command (no alias and not in menus or toolbars) allows clipping images. By default, the command prompts for the selection of one or more attached images, a boundary definition, rectangular, and the two corners of the rectangular boundary.

```
: CLIPIMAGE
Specify image to clip» Selection
Default: Create boundary
Options: ON, OFf, Delete or Create boundary
Specify clipping option» Enter
Default: Rectangular
Options: Polygonal, Rectangular or Select polyline
Specify boundary option» Enter
Specify first corner point» P1
Specify opposite corner point» P2
```

Existing options are similar to the options of the CLIPREFERENCE command.

Advanced layer commands

Real projects usually include a large amount of layers. Managing these layers can be quite time consuming. DraftSight includes the following commands; most of them allow managing layers by entities selection.

- The HIDELAYER command (no alias, 🖫 on the **Layer Tools** toolbar, or **Format | Layer Tools | Hide Layer** main menu) allows turning layers off by selecting entities that belong to those layers.

- The SHOWLAYERS command (no alias, 🖫 on the **Layer Tools** toolbar, or **Format | Layer Tools | Show All Layers** main menu) turns on all layers. Only frozen layers remain invisible.

- The FREEZELAYER command (no alias, 🔲 on the **Layer Tools** toolbar, or **Format | Layer Tools | Freeze Layer** main menu) allows the user to freeze layers by selecting entities that belong to those layers. The current layer cannot be frozen.

- The THAWLAYERS command (no alias 🔲 on the **Layer Tools** toolbar, or **Format | Layer Tools | Thaw All Layers** main menu) thaws all layers.

- The LOCKLAYER command (no alias, 🔲 on the **Layer Tools** toolbar, or **Format | Layer Tools | Lock Layer** main menu) allows the user to lock a layer by selecting one entity that belongs to that layer. Unlike previous commands, only one entity can be selected.

- The UNLOCKLAYER command (no alias, 🔲 on the **Layer Tools** toolbar, or **Format | Layer Tools | Unlock Layer** main menu) allows the user to unlock a layer by selecting one entity that belongs to that layer. Unlike previous commands, only one entity can be selected.

- The ISOLATELAYER command (no alias, 🔲 on the **Layer Tools** toolbar, or **Format | Layer Tools | Isolate Layer** main menu) allows the user to isolate layers by selecting entities that belong to those layers. Only the layers from selected entities remain visible, all others are turned off.

- The UNISOLATELAYER command (no alias, 🔲 on the **Layer Tools** toolbar, or **Format | Layer Tools | Unisolate Layer** main menu) turns on layers that were turned off by the ISOLATELAYER command.

- The ACTIVATELAYER command (no alias, ✓ on the **Layer Tools** toolbar, or **Format | Layer Tools | Activate Layer** main menu) activates a layer by selecting an entity belonging to that layer.

- The TOACTIVELAYER command (no alias, 🔲 on the **Layer Tools** toolbar, or **Format | Layer Tools | Entity to Active Layer** main menu) modifies the layer of selected entities to the current layer.

- The MATCHLAYER command (no alias, 🔲 on the **Layer Tools** toolbar, or **Format | Layer Tools | Change Entity's Layer** main menu) modifies the layer of selected entities to the layer of a destination entity to be selected.

- The DELETELAYER command (no alias, 🔲 on the **Layer Tools** toolbar, or **Format | Layer Tools | Delete** main menu) erases a layer definition, including all entities belonging to it. It is possible to select one entity belonging to that layer, or type its name, if no entity is visible.

- The UNDOLAYER command (no alias, 🔲 on the **Layer Tools** toolbar, or **Format | Layer Tools | Restore Layer's State** main menu) undoes the last modification of layers properties. It can be used in sequence and does not affect entities creation.

Exercise 11.1

With the project being developed along the book, we are going to apply some of these advanced layer commands:

1. Open the drawing PROJECT1-CH10.DWG.

2. Activate the **Model** space.

3. Apply the LOCKLAYER command and select one of the visible parts edges. The **OUTLINE** layer has been locked. If we try to erase or modify entities in this layer, we get the information that these belong to a locked layer.

4. Apply the ACTIVATELAYER command and select the **SECTION AA** text. The **CUTTING TEXT** layer becomes current.

5. Apply the FREEZELAYER command and select one of the dimensions and the table. As we select entities, their layers are being frozen. So, **DIMENSION** and **TABLE** layers are frozen.

6. Unlock layer of the parts edges by applying the UNLOCKLAYER command.

7. Finally, apply the ISOLATELAYER command and select one of the parts edges. The **OUTLINE** layer becomes current and all others are turned off.

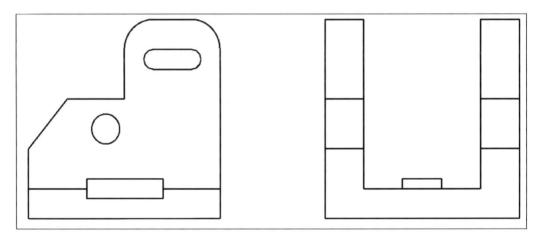

8. There is no need to save the drawing.

Summary

This last chapter includes how to reference other drawings and images in our drawing and several additional commands to deal with layers. When opening a drawing with references, these are loaded at the current state and cannot be directly modified.

The REFERENCES command displays a palette to manage referenced external drawings and images. The CLIPREFERENCE command allows the user to clip referenced drawings or blocks. The CLIPIMAGE command allows the user to clip images.

The HIDELAYER command allows the user to turn layers off by selecting entities that belong to those layers. The SHOWLAYERS command turns on all layers. The FREEZELAYER command allows the user to freeze layers by selecting entities that belong to those layers. The THAWLAYERS command thaws all layers. The LOCKLAYER command allows the user to lock a layer by selecting one entity that belongs to that layer. The UNLOCKLAYER command allows the user to unlock a layer by selecting one entity that belongs to that layer. The ISOLATELAYER command allows the user to isolate layers by selecting entities that belong to those layers. The UNISOLATELAYER command turns on layers that were turned off by the ISOLATELAYER command. The ACTIVATELAYER command activates a layer by selecting an entity belonging to that layer. The TOACTIVELAYER command modifies the layer of selected entities to the current layer. The MATCHLAYER command modifies the layer of selected entities to the layer of a destination entity to be selected. The DELETELAYER command erases a layer definition, including all entities belonging to it. The UNDOLAYER command undoes last modification of layers properties.

Index

Symbols

Freehand option 85
FREEZELAYER command 153

G

GETAREA command   64
GETDISTANCE command 64
GETPROPERTIES command 67
GETXY command 66
Global scale textbox 59
gradients
  applying 109, 110
Graebert GmbH
  URL 7
graphical auxiliary tools
  about 22
  entity snaps 24
  entity snaps, tracking 25
  Grid function 26
  Ortho function 22
  Polar function 22, 23
  Snap function 26
  visualization tools 26, 27
Grid function 26

H

H. *See* HATCH command
Halfwidth option 80
HATCH command 110, 106
Hatch dialog box 108
HE. *See* EDITHATCH command
HIDELAYER command 152
Highlight boundary entities 108
home palette 7
Horizontal and Vertical option 114

I

I. *See* INSERTBLOCK command
ID. *See* GETXY command
image
  attaching 151
  external images, clipping 151
  external images, linking 148
  external images, managing 148
  external images, referencing 147

Incremental angles for Polar guide display
    list 23
Increment option 88
INSERTBLOCK command 94, 99
Insertion method area 86
Insertion point 96, 98
ISO 13567 (Europe) 55
ISOLATELAYER command 153

J

J. *See* WELD command
Join option 80

L

L. *See* LINE command
LA. *See* LAYER command
Last method 35
LAYER command 53
layer list, columns
  description 55
  frozen 54
  lock 54
  name 54
  Print 55
  PrintStyle 55
  show 54
  status 54
layers
  about 51
  activating 52
  colors, applying 56, 57
  entities, modifying 52
  filtering 55
  freezing 53
  linestyles, applying 57, 59
  lineweights, applying 59, 60
  locking 53
  thawing 53
  turning, off 52
  turning, on 52
  unlocking 53
Layers Manager dialog box 57
layouts 131
LE. *See* SMARTLEADER command

W

W. *See* **EXPORTDRAWING command**
WELD command 87
Width option 80
WPolygon method 36

X

X. *See* **EXPLODE command**
XA. *See* **ATTACHDRAWING command**
XC. *See* **CLIPREFERENCE command**

Z

ZB. *See* **ZOOMBACK**
ZOOMBACK 27
ZOOM command 27
zoom fit 26
zoom in/zoom out 26

Thank you for buying
Getting Started with DraftSight

About Packt Publishing

Packt, pronounced 'packed', published its first book "*Mastering phpMyAdmin for Effective MySQL Management*" in April 2004 and subsequently continued to specialize in publishing highly focused books on specific technologies and solutions.

Our books and publications share the experiences of your fellow IT professionals in adapting and customizing today's systems, applications, and frameworks. Our solution based books give you the knowledge and power to customize the software and technologies you're using to get the job done. Packt books are more specific and less general than the IT books you have seen in the past. Our unique business model allows us to bring you more focused information, giving you more of what you need to know, and less of what you don't.

Packt is a modern, yet unique publishing company, which focuses on producing quality, cutting-edge books for communities of developers, administrators, and newbies alike. For more information, please visit our website: www.packtpub.com.

Writing for Packt

We welcome all inquiries from people who are interested in authoring. Book proposals should be sent to author@packtpub.com. If your book idea is still at an early stage and you would like to discuss it first before writing a formal book proposal, contact us; one of our commissioning editors will get in touch with you.

We're not just looking for published authors; if you have strong technical skills but no writing experience, our experienced editors can help you develop a writing career, or simply get some additional reward for your expertise.

FreeCAD [How-to]

ISBN: 978-1-84951-886-4   Paperback: 70 pages

Solid Modeling with the power of Python

1. Learn something new in an Instant! A short, fast, focused guide delivering immediate results.

2. Packed with simple and interesting examples of python coding for the CAD world.

3. Understand FreeCAD's approach to modeling and see how Python puts unprecedented power in the hands of users.

4. Dive into FreeCAD and its underlying scripting language.

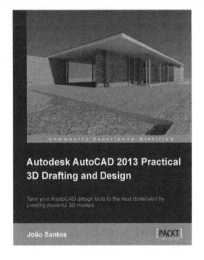

Autodesk AutoCAD 2013 Practical 3D Drafting and Design

ISBN: 978-1-84969-935-8   Paperback: 374 pages

Take your AuotoCAD design skills to the next dimension by creating powerful 3D models

1. Obtain 2D drawings from 3D models

2. Master AutoCAD's third dimension

3. Full of practical tips and examples to help take your skills to the next dimension

Please check **www.PacktPub.com** for information on our titles

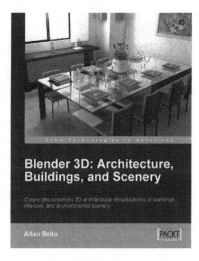

Blender 3D: Architecture, Buildings, and Scenery

ISBN: 978-1-84719-367-4   Paperback: 332 pages

Create photorealistic 3D architectural visualizations of buildings, interiors, and environmental scenery

1. Turn your architectural plans into a model

2. Study modeling, materials, textures, and light basics in Blender

3. Create photo-realistic images in detail

4. Create realistic virtual tours of buildings and scenes

Adobe Flash 11 Stage3D (Molehill) Game Programming Beginner's Guide

ISBN: 978-1-84969-168-0   Paperback: 412 pages

A step-by-step tutorial for creating custom Facebook applications using the Facebook platform and PHP

1. The first book on Adobe's Flash 11 Stage3D, previously codenamed Molehill

2. Build hardware-accelerated 3D games with a blazingly fast frame rate.

3. Full of screenshots and ActionScript 3 source code, each chapter builds upon a real-world example game project step-by-step.

Please check **www.PacktPub.com** for information on our titles

Printed in Great Britain
by Amazon